get book
p. 28

R. K. Hauser

The sea breeze affects our lives in many ways, it controls our local weather, not only on the coast but also in many districts inland. Air pollution and smog, the distribution of airborne insect pests and the spread of pollen are all controlled by the sea breeze. In the world of sport it is important to glider pilots, sailors and surfers, and balloonists.

In this book we see how radar, lidar and satellite photography have helped to forecast and map sea breeze and the all-important 'sea-breeze front'. The book ends with a description of laboratory experiments, mostly carried out by the author and his co-workers, and a simple summary of theoretical models. This book will be welcomed by those researching in the subject but will also be valuable to the general reader who is interested in local weather and the natural environment.

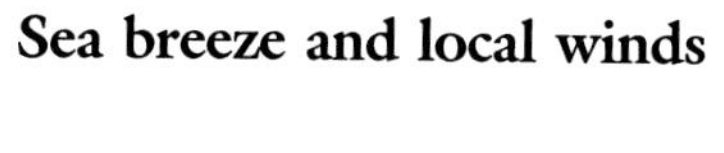

Sea breeze and local winds

Sea breeze and local winds

JOHN E. SIMPSON
Department of Applied Mathematics and Theoretical Physics, University of Cambridge

CAMBRIDGE
UNIVERSITY PRESS

Published by the Press Syndicate of the University of Cambridge
The Pitt Building, Trumpington Street, Cambridge CB2 1RP
40 West 20th Street, New York, NY 10011-4211, USA
10 Stamford Road, Oakleigh, Melbourne 3166, Australia

First published 1994

Printed in Great Britain at The University Press, Cambridge

A catalogue record for this book is available from the British Library

Library of Congress cataloguing in publication data

Simpson, John E., 1915–
Sea breeze and local winds / John E. Simpson.
p. cm.
Includes bibliographical references and index.
ISBN 0 521 45211 2
1. Sea breeze. I. Title
QC939.L37S56 1994
551.5′185–dc20 93-29979 CIP

ISBN 0 521 45211 2 hardback

wv

Contents

Foreword

Everyone depends on the atmosphere for their breathing, for its protection of the earth's surface from the solar radiation, its winds for ventilation and transport, its rain which feeds the crops and its clouds which help prevent us freezing or roasting. Of course there have been many books ranging from the most popular to the most learned on the atmosphere and its motion. But few books have focussed on the special features of the atmosphere caused by the effects of the sea on the climate and weather of land areas near the coasts. Since most people in the world live within 200 km of the coasts, it is not surprising that this aspect of meteorology has been important since the time of the ancient Greeks.

John Simpson has always been an active meteorologist, during his career as a glider pilot, a science teacher and latterly a university research scientist, and an excellent photographer. I have known him as a colleague at Cambridge over the past 17 years and was guided by him to see the atmosphere with new eyes. Firstly we all learnt, after he arrived in Cambridge, that indeed sea breezes reached us at about 7pm (7 arrived in the remarkable summer of 1976). He had set up a set of stations between us and the coast to track its movement. He also began to interest a number of colleagues to join with him in a series of laboratory experiments that quite changed the scientific understanding of 'gravity currents', which are the basic mechanism for driving the cool sea breezes inland. He has received visitors from far and wide and had an enormous correspondence; there is no sea breeze in the world that he has not heard about. This book is an excellent account of them; their history, their different types depending on the coastline or the synoptic situation, their connection with local weather such as clouds and rain, their effects on pollution, aircraft and bird flight, their measurements which nowadays includes radar, and finally some of the laboratory studies at Cambridge and elsewhere.

Professor Julian Hunt, FRS
Chief Executive, Meteorological Office, Bracknell

Preface

I first met the sea breeze over 50 years ago when, due to my ignorance of its existence, I landed a glider downwind after a long flight to the east coast. The result of this ignorance was nothing worse than an over-shoot into a field of beans.

My later experience of the sea breeze was obtained in gliders, finding out the nature of the sea-breeze front. This led to a career in Applied Mathematics developing an interest in the physics of the atmospheric boundary layer, backed up with laboratory experiments. In all this, aspects of the sea breeze kept turning up.

The first nine chapters of the book are aimed at the general reader; they deal with the behaviour of the sea breeze and details which can be seen both from the ground and from the air. Other local winds, some of which are closely related, are also dealt with in this section. It is shown that the sea breeze affects the lives of humans in many ways, for example in the distribution of pollution and in the ways in which people spend their leisure time. The last three chapters are slightly more technical and deal with measurements of sea-breeze phenomena.

I am grateful to Professor Julian Hunt for writing the Foreword to this book and for the help of Professors R.S. Scorer, Herbert Huppert and to Drs. Anthony Edwards and John Chapman for reading the text and for their helpful comments. My thanks go to Margaret Downing for her able preparation of the diagrams and to Jason Newling for photographic work.

Ken Griffin's water colour of the 'sea breeze tree' on the coast of Majorca inspired the design for the cover of the book.

1

The sea breeze

1.1 Introduction

The 'sea breeze', which flows inland at the coastline on fine days, is caused by the temperature difference between the hot land and the cool sea. This difference increases during the day and produces a pressure difference at low levels in the atmosphere, which causes the low-level sea-breeze to blow. At night this pressure difference disappears and is sometimes reversed, causing a 'land breeze'. Of these daily alternating winds the sea breeze is the much stronger effect.

At the coast of many hot tropical countries, where the overall pressure gradient is steady from day to day, the arrival of the sea breeze can be expected regularly every day at the same time, reaching a strength of 6 or 7 m s^{-1}. In temperate climates the sea breeze also blows on sunny days, but winds from different directions caused by the movement of depressions and anticyclones often modify its development. The sea breeze will start to blow when the temperature difference between the land and sea is large enough to overcome any offshore wind. For example, on the coast of southern England on a calm day a temperature difference of 1 °C is large enough for a sea breeze to form, but to overcome an offshore wind as strong as 8 m s^{-1} a temperature difference of 11 °C is needed (Watts, 1955).

Early in the day, soon after its onset, the depth of the sea breeze may be less than 50 m and the wind just above it can be blowing in the opposite direction. This is shown clearly in figure 1.1, where all the plume from the lower chimney (75 m) is contained in the sea breeze. The plume from the taller chimney, (150 m) all moves in the opposite direction to the low-level breeze. As the day develops, so does the thickness of the sea breeze, often reaching a depth of 300 m. Later in the day the direction of the wind shifts a little due to the Earth's rotation, veering a few degrees in the Northern Hemisphere.

Figure 1.1. The depth of the sea breeze illustrated by smoke plumes. The plume from the lower chimney, height 75 m, is all contained in the low-level breeze; that from the taller chimney all moves in the opposite direction. (Photo by Ralph Turncote.)

Fresh sea breezes appear to be a pleasant feature of life near the coast, however they may have harmful effects on the distribution of pollution. In California, for example, the sea breeze generally has a 'purging' effect, but the stable layering together with the diurnal reversal of both the mountain–valley and land–sea breeze are a problem, maintaining dangerous concentrations of pollution at Los Angeles and other towns on the coast. The city of Athens gives another important example of the daily return of pollution each afternoon.

Sea-breeze 'smog' is also a feature of some parts of England, where the arrival of the sea breeze may appear as a wall of smoke, as shown in figure 1.2, which was taken from the air near Middlesborough (Eggleton & Atkins, 1972).

Airborne insect pests such as locusts and aphids find the sea breeze useful since it may protect them from being blown out to sea on days when the overall wind is towards the ocean.

The developing sea breeze in calm weather gradually extends further out to sea as well as inland. On days with an offshore wind the sea breeze may also eventually spread inland, but this will happen later during the day and may be associated with a sudden squall. In an early account of the sea breeze at Cohasset, Massachusetts, Appleton, (1892) described how white-caps were seen on the sea as the interaction between the prevailing winds and the sea breeze produced a sharp sea-breeze front, moving slowly towards the land.

At about the same time in history the arrival and penetration of the sea breeze into the coastal area near Boston was investigated by a group from Harvard College Observatory (Davis, Schultz & Ward, 1890). Over 100 people took

Figure 1.2. The advancing front of the sea breeze, seen as a wall of smoke. Taken from a glider, south-west of Middlesborough, 8 August 1966. (Photo by M. Randle.)

part in making observations over a total period of three months, and the advance of the sea breeze inland was tracked on 30 occasions. This was a remarkable undertaking when we bear in mind that it had to be carried out without the use of telephones or automobiles. Figure 1.3 is an example of the results, showing the inland penetration of the sea breeze on 26 July 1887.

In hot countries the arrival of the sea breeze is very welcome as a gust of cooling wind in the hottest part of the day. In parts of Australia it is known as 'the Doctor'. In West Pakistan, for example, the sea breeze is well known locally at points inland for its tempering effect on the fierce summer heat. At Hyderabad, in Sind, 170 km inland, every house has a 'wind catcher' – a wooden tunnel built above the roof-tops to channel the cool sea breeze, which arrives in the evening, into the rooms below. These structures and other examples of wind catchers are described in Chapter 8.

1.2 Sea-breeze clouds

The convergence of the winds near the sea-breeze boundary must cause air to rise, which often condenses and forms clouds. The extent of the sea breeze can sometimes be deduced from the presence of distinctive clouds which form in this zone. A line of cloud parallel to the coast on an otherwise cloudless day is a clear sign of the boundary of the sea breeze, the so-called 'sea-breeze front'.

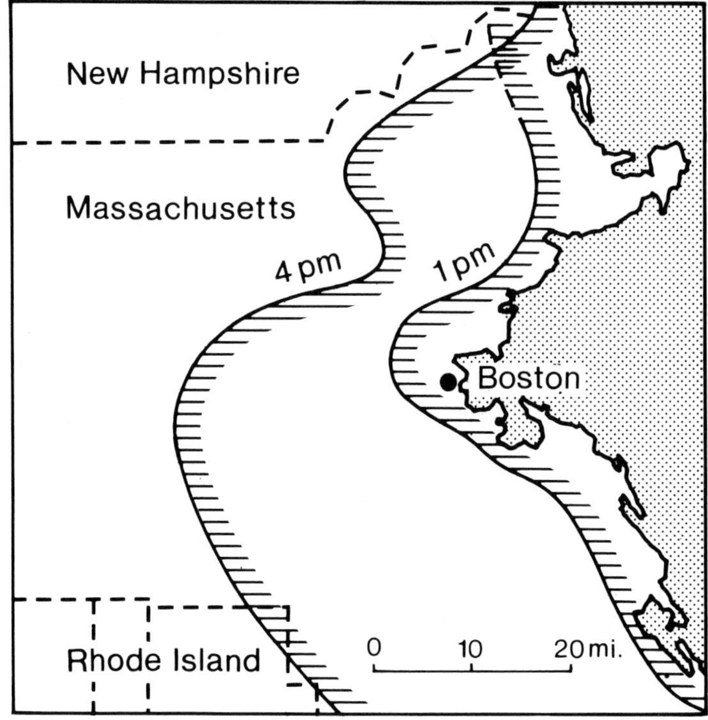

Figure 1.3. The boundary of the sea breeze, measured at 1300 h and 1600 h on 26 July 1887. (After Davis, Schultz & Ward, 1890.)

The interpretation of cloud patterns often makes it possible to see how far the sea breeze has spread inland. An example of this is given in figure 1.4, a view from the air soon after taking off from Gatwick in southern England, just before reaching the south coast. On the left the sky is full of small cumulus clouds uniformly spaced above the heated ground and indicating thermals of rising air. The sea is just out of sight on the right, and the air in the sea breeze is cloudless, but at the weak convergence zone between the land- and sea-air a line of larger cumulus clouds can be seen.

In different parts of the world, for various reasons, these convergence lines can be very intense and develop large banks of clouds, causing rain and even thunderstorms. In such localities the presence of the sea breeze has a marked effect on the climate.

1.3 The sea breeze in history

The Greeks did not like being on board ship at night, but nevertheless they would set sail after sunset to take advantage of the land breeze. This was also

Figure 1.4. The boundary of the sea breeze, which is blowing from the right, is marked by the edge of the cumulus clouds. Seen from the air just south of Gatwick. (Photograph courtesy of Colin Street.)

the custom of many fishermen of the Greek islands who would later use the sea breeze to return to port in the morning.

An outstanding case in Greek history of the use of the sea breeze was by the Athenian leader Themistocles, the commander of the Greek forces at the naval battle of Salamis in 480BC, where both the place and time for the battle were wisely chosen (Plutach, transl. 1892). The place was the channel between the island of Salamis and the mainland, the time chosen was when a brisk wind would start to blow from the open sea (a sea breeze) and raise waves in the narrow channel. The rough water did not inconvenience the Greek ships, which were solidly constructed and lay low in the water. The Persian ships, with lofty sterns and decks, were clumsy and unwieldy and managed poorly in high waves. Therefore, when the wind reached a fair strength the Greek commander ordered attack and the Persian fleet were shattered in the ensuing battle.

Another military use of the sea breeze is recorded about 2000 years later during the American Civil War when balloons were used for aerial reconnaissance (Haydon, 1941). John La Fountain was the first man with the Union

Army to make free reconnaissance flights over enemy territory. He would drift eastwards across the enemy lines when the sea breeze near ground level favoured him, then discharging ballast he relied on the prevailing westerly air stream at higher levels to carry him back. He repeated this feat many times, but had a narrow escape when Union troops seeing a balloon coming from Confederate territory took him for an enemy.

1.4 The onset of the sea breeze

As a summary of the general features of the sea breeze we cannot do better than quote this description by the sea-captain William Dampier from his *Voyages*

> These sea breezes do commonly rise in the Morning about Nine-a-Clock, sometimes sooner, sometimes later: they first approach the Shore so gently, as if they were afraid to corne near it, and oft-times they make some faint Breathings, and as if not willing to offend, they make a halt, and seem ready to retire. I have waited many a time both Ashore to receive the Pleasure, and at Sea to take the Benefit of it.
>
> It comes in a fine, small, black Curl upon the Water, when, as all the Sea between it and the Shore not yet reached by it, is as smooth and even as Glass in comparison; in half an Hour's time after it has reached the Shore it fans pretty briskly, and so increaseth gradually till Twelve a-Clock, then it is commonly strongest, and lasts so till Two or Three a very brisk Gale; about Twelve at Noon it also veers off to Sea Two or Three points, or more in very fine Weather. After Three a-Clock it begins to die away, and gradually withdraws its force till all is spent, and about Five a-Clock, sooner or later, according to the Weather is, it is lull'd asleep, and comes no more till the next morning.
>
> Land breezes are quite contrary to the sea-breezes; for these blow right from the shore, and as sea breezes do blow during the day and rest during the night; so on the contrary, these do blow in the night and rest during the day, and so they do alternately succeed each other.
>
> (Dampier, 1670)

William Dampier gave a clear account of the sea breeze from the point of view of the professional sailor; slightly different stories might be expected from an experienced air-pilot or from a student of air pollution. Much of their knowledge could be valuable for a weather-wise person only interested in making the best of a sea-side holiday.

2

Formation of the sea breeze

2.1 Land and sea-breeze generation

Aristotle believed that wind was a 'dry exhalation' and because no such exhalation could be expected to originate from the damp sea, he had difficulty in explaining the generation of the sea breeze without the idea of the rebounding of land breezes at obstacles. The need to introduce the complicated factors of obstacles and reflux is reminiscent of the need met at a later date by the astronomer Ptolemy for introducing epicycles to make his geocentric system of the planets fit the observations.

More convincing to the modern mind, but still not entirely satisfactory, is the simple picture given in many school geography books of the air rising above the heated land, leaving a gap to be filled in by the inflowing sea breeze.

Figure 2.1 gives the generally accepted explanation of the development of the pressure field which gives rise to the low-level sea breeze. When the sun shines, the sea surface temperature changes very little, but the land becomes hotter and convection currents of air distribute heat through several thousand feet above the ground. No changes occur above a certain height, so the sideways expansion of each column of air above the land, B, produces changes in pressure which are transmitted sideways with the speed of sound. The resulting pressure difference at low levels is responsible for the onset of the sea breeze.

A weaker return flow aloft is necessary to balance the system.

2.2 Pressure patterns and the sea breeze

The growth and extent of the pressure field at any point is of primary importance as it supplies the driving force for the sea breeze. If barometer readings are carefully examined in very calm weather, when no appreciable synoptic changes

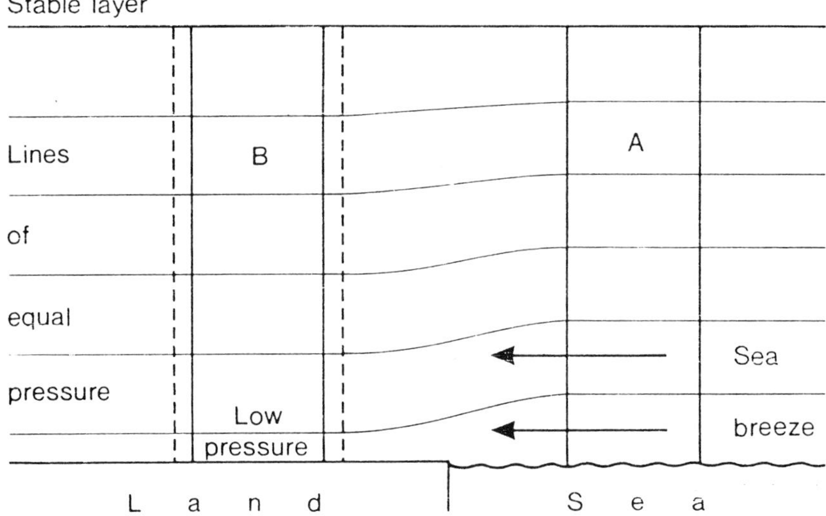

Figure 2.1. Development of the pressure field which gives rise to the sea breeze at lower levels. A column of air above the land, B, is heated by the sun and must expand sideways, as shown by the dashed lines. A column above the sea, A, is unaltered. This causes a pressure difference at low levels which gives rise to the sea breeze.

are taking place, a daily variation of surface barometric pressure can be detected. If averages are taken over a period of several hundred days at any particular place it becomes clear that a regular 'atmospheric tide' exists.

This diurnal march of the barometer can be considered as the result of two waves of different origin and character.

One of these, which is a semi-diurnal wave, is analogous to the waves in the ocean produced by the attraction of the sun and moon. In the sea the moon's tidal power is 2.4 times that of the sun but in the atmosphere the sun-tide is 15 to 20 times as strong as the moon-tide.

The second wave is different from the first wave since it has a period of one day and does not depend, like the first wave, only on the latitude and the season. This wave is produced by the variation of temperature in the lower layers of the atmosphere, and is called the thermal wave.

The thermal wave is influenced by the difference in the diurnal variations between land and sea. This difference produces the land- and sea-breeze phenomenon, and brings corresponding variations in the form of the thermal wave.

These waves are illustrated in figure 2.2. Figure 2.2(*a*) shows the diurnal variation in pressure anomaly at two sites. The total variation is seen to be semi-diurnal and is recorded at the surface both at Jersey in the Channel Islands and at Paris. Figure 2.2(*b*) shows the difference between the pressure at Jersey,

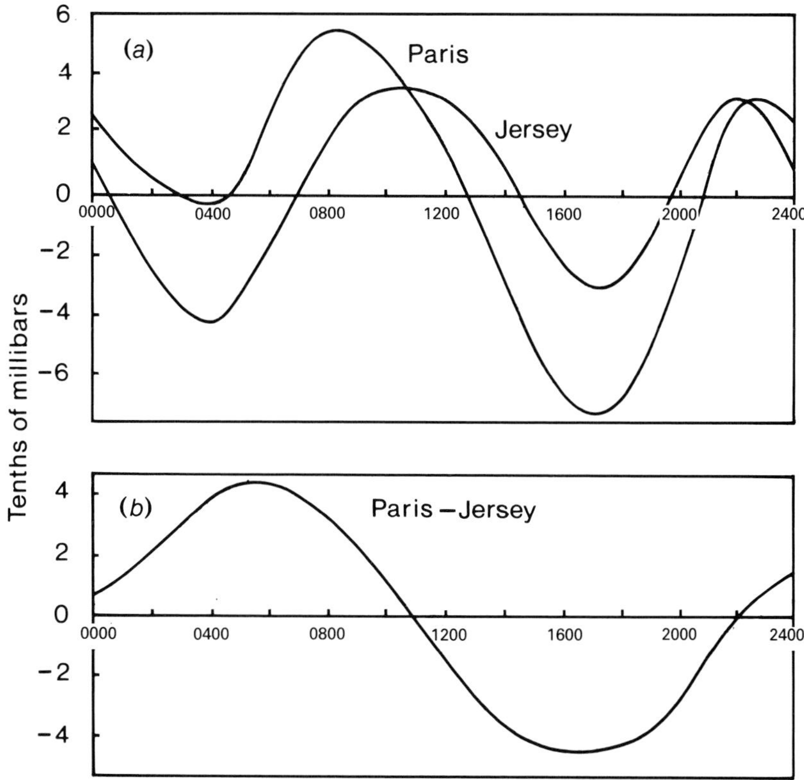

Figure 2.2. (*a*) Departures from the mean surface pressure in June at Jersey and Paris. Semi-diurnal pressure waves, or aerial tides. (*b*) Pressure differences between Jersey and Paris. This is a single diurnal wave, related to surface heating.

an oceanic site, and at Paris, which is 160 km inland. The difference is a single diurnal wave.

The spacial variation of the size of this thermal wave across Europe is illustrated in figure 2.3, which shows how the surface pressure changes during a morning in June (Met. Office, 1943). This map shows pressure tendencies at 1300 GMT, i.e. the change between 1000 and 1300 GMT. Note the large negative tendency over Spain and Central Europe due to the divergence of the air above in the middle of the day and even a centre of negative tendency over the British Isles. It can be seen that the tendency at Jersey is 0, and that at Paris is −4, a difference of about four tenths of a millibar, agreeing with the results of figure 2.2.

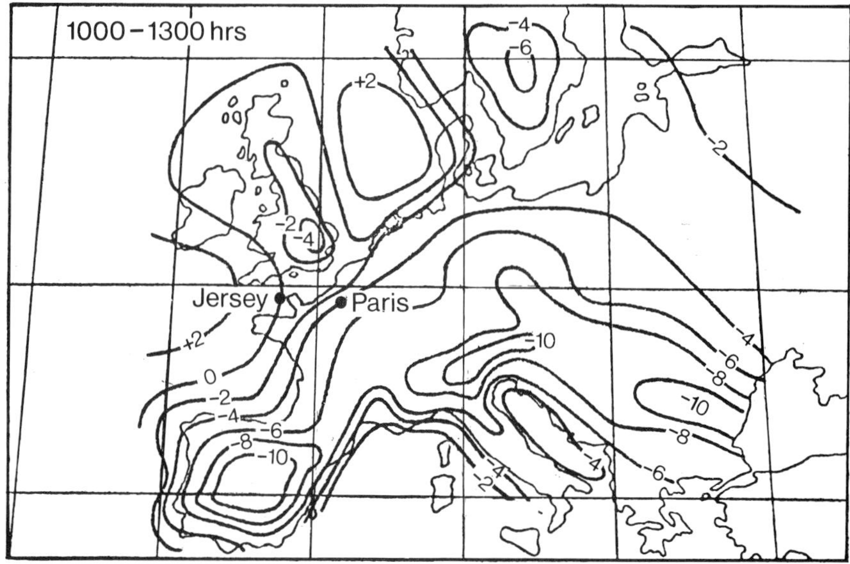

Figure 2.3. Surface pressure changes in Europe between 1000 and 1300 GMT in June. Units are tenths of a millibar; the positions of Jersey and Paris are marked. (After Met. Office 1943, unpublished.)

2.21 Diurnal temperature and pressure changes

The sea breeze does not necessarily depend on high surface temperature since it is the changes in temperature which are important.

Sometimes these temperature changes may be very great, for example tests made in Arizona have shown that during the day black asphalt reaches a temperature 19 °C above normal surroundings. It has been suggested that a large area of black asphalt near the coast would induce a sea-breeze circulation, leading to cloud formation and rain (Black & Tarny, 1963). Calculations suggest that the optimum length would be 50 kilometres inland from the shore, with a width $\frac{1}{10}$ to $\frac{1}{5}$ of the length. One acre of asphalt should be sufficient for three acres of arable land. Suitable large-scale test sites would be in Libya, Venezuela or W. Australia.

In countries such as Britain the daily temperature changes are much less than this. Typical mean temperature changes here in June between 0900 and 1500 GMT amount to only about 2 °C towards the centre of the country.

The pressure drop during the same time shows very similar contour lines to those of temperature rise, with the minima inland.

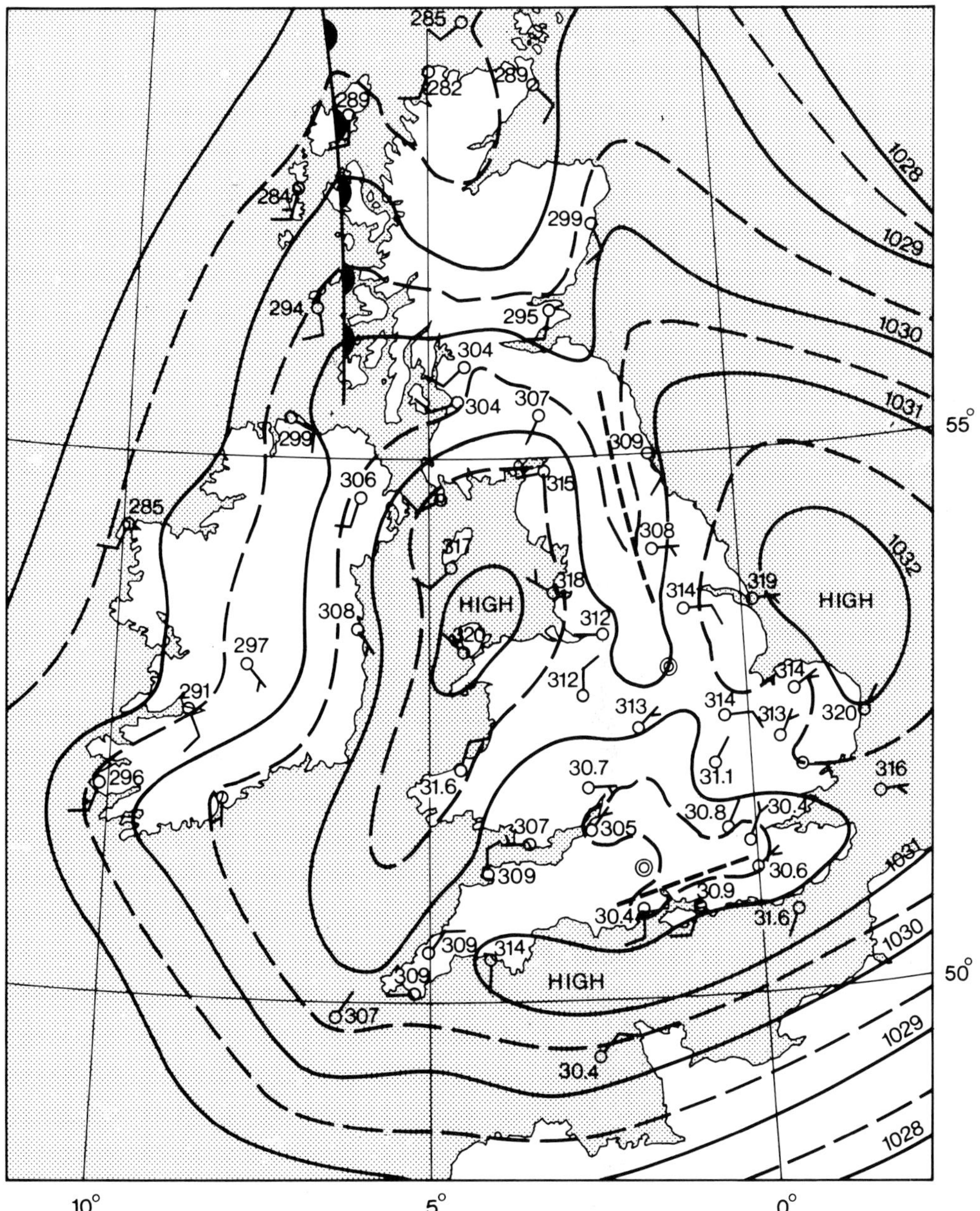

Figure 2.4. Detailed surface pressure field at 1800 GMT, 26 July 1963, a day with strong inland sea-breeze penetration. The two dashed lines, near the north-east and the south coast, show the observed position of the sea-breeze front.

2.22 *Pressure field on a day of strong sea breeze*

In the course of some sea-breeze investigations in the south of England (Simpson, 1964) numerous charts of the developing pressure field were plotted. Using the hourly charts of the Meteorological Office it was possible to see the development of the pressure pattern not only over the land but also over the sea. One of these maps of pressure contours, made on 26 July 1963, a good sea-breeze day, is shown in figure 2.4. Areas of high pressure appear above the North Sea, English Channel and Irish Sea. The two dashed lines, one near the North East coast and the other in the South, show the position of the boundary of the sea breeze (the sea-breeze front) at 1800 GMT.

2.23 *Seasonal differences of temperature*

Monthly mean air temperatures and corresponding sea-surface temperatures have been measured in order to examine the probable differences available to drive the sea breeze. Sea temperatures are difficult to measure reliably, but some good long-term station records are available from the US Dept. of Commerce. Figure 2.5 shows the sea-surface temperature curve lagging behind that of the air temperature by about a month, as is normally expected. (Diaz & Quayle, 1980).

2.3 Sea-breeze strength and direction: hodographs

A useful way of presenting measurements of the sea breeze is to use a wind hodograph traced out by the end point of the wind vector as its value changes with time. Hodographs show, at any given locality, a complete 360° turn during the day, provided the overall gradient wind is weak.

In addition to many observational results, there have been theoretical investigations of hodographs in the land–sea-breeze circulation. Pioneering work (Haurwitz, 1947) in a simplified linear model considering friction, Coriolis forces due to the Earth's rotation and inertial forces showed that the hodographs should be ellipses for which rotation is always clockwise (in the northern hemisphere). In this model the sea breeze, which starts blowing at right angles to the shore, gradually veers in direction until by sunset it is blowing nearly parallel to the coastline.

Sea-breeze hodographs measured in the northern hemisphere mostly show a clockwise rotation with time, but there are also stations with a clear anticlockwise direction. Theory suggests that the rotation displayed in hodographs is

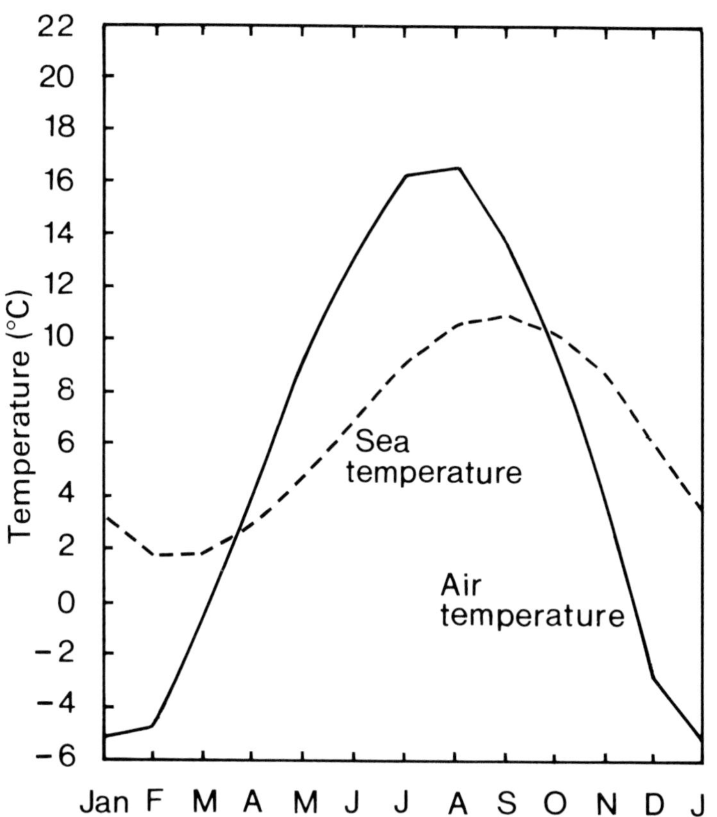

Figure 2.5. To show the difference between land and sea temperature throughout the year, measured at Eastport, Maine. (Average through the years 1927–77.) (After Diaz & Quayle, 1980.)

primarily due to the Earth's rotation, but some other explanation must be found for those with anticlockwise rotation. In the following sections complex coastlines and mountains are shown to be some of the causes of anticlockwise rotation.

2.31 Observed diurnal changes in sea-breeze direction

The original simple theoretical model was for a straight coastline with uniform conditions inland; in reality, the coastline may not be straight and there may be significant features which can affect the growth of the pressure field which creates the sea breeze.

The combination of two sea breezes originating at two coastlines which enclose an angle to 70 at Halifax, Nova Scotia, was found to be different from that to be expected on a straight coast (Dexter, 1958). The graphical addition of the two corresponding elliptical hodographs was shown to produce a hodograph very similar to that observed; see figure 2.6.

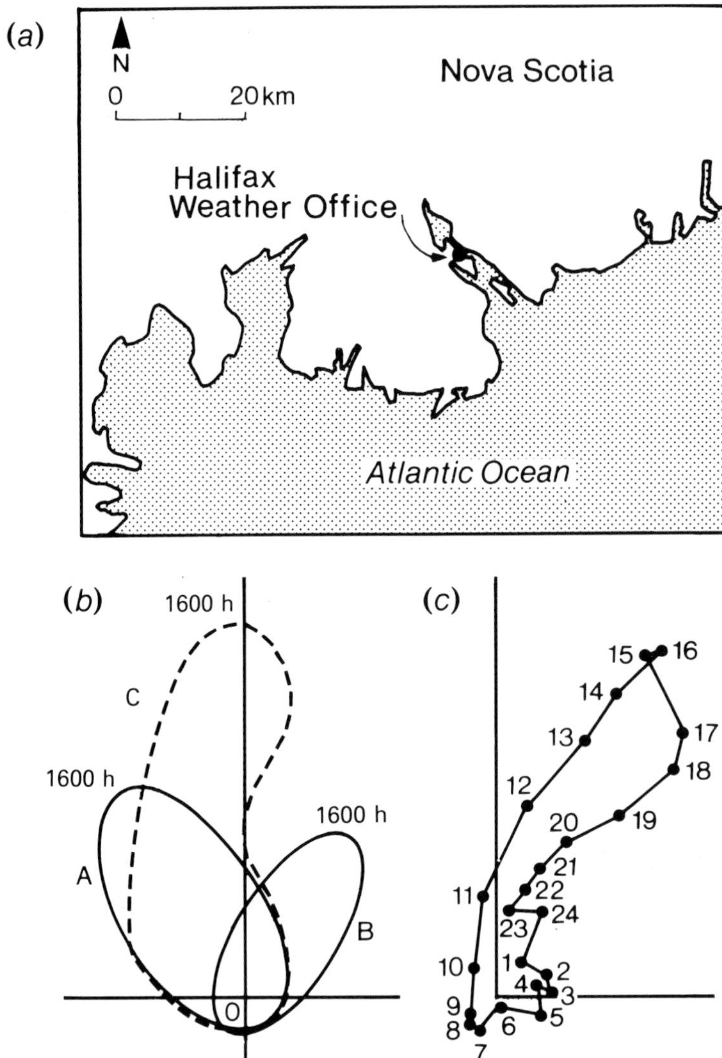

Figure 2.6. The harbour sea breeze and the mainland sea breeze near Halifax, Novia Scotia. (*a*) Map of the area. (*b*) Theoretical sea-breeze hodographs. A, Halifax harbour effect; B, the Atlantic effect; C, the sum of A and B, allowing for the periods in which they apply. For example, the wind was nearly from the south at 1600 h. (*c*) Measured hodograph for June, showing a similar shape to C. The numbers 1–24 represent the times 0100–0000 h. (After Dexter, 1958.)

Deviations from a standard hodograph were found during observations of the sea breeze in a Gulf in South Australia (Physick & Byron-Scott, 1977). The map in figure 2.7 shows St. Vincents Gulf, which is about 60 km wide in the southerly parts. The study here was concerned with the effect of the sea breeze on the airborne pollution from the city of Adelaide, which lies on the East coast of the Gulf. Winds were measured on the west coast of the Gulf and across to the east coast.

The development of the sea breeze was explained in terms of a 'local gulf sea breeze' and a 'continental sea breeze'. The shorelines of the Gulf experience two separate sea-breeze systems. During the morning and early afternoon sea breezes blow from opposite directions, basically due to the temperature difference between the land and gulf waters, occurring at respective shorelines. At some stage during the afternoon a continental sea breeze, basically due to the temperature difference between the Australian continent and the Southern Ocean, arrived from a southerly direction and superposed itself on the Gulf sea breeze.

Figure 2.7 illustrates the way in which the sea breeze could be compounded to form the observed wind on each shoreline. This effect may often be the explanation for the difference of observed hodographs from the clockwise ellipse; another cause may be the effect of mountains.

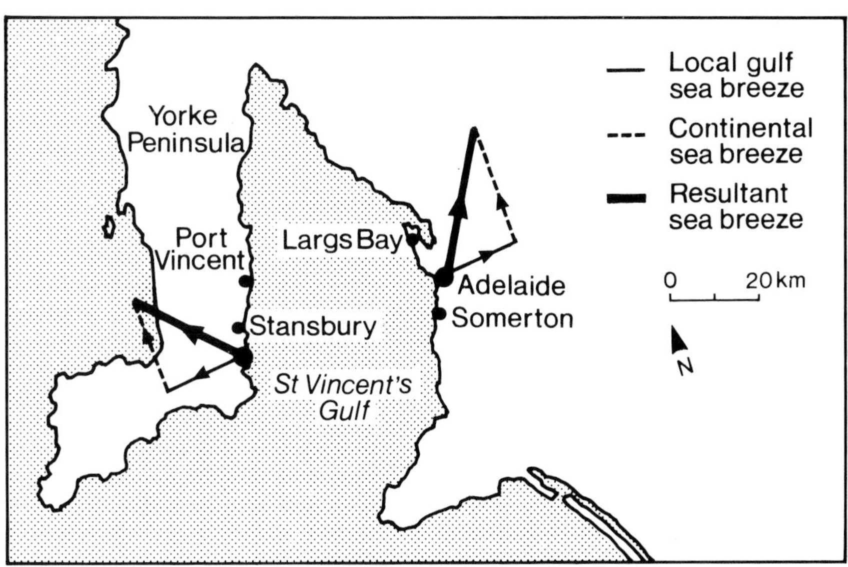

Figure 2.7. Effect of St. Vincent's Gulf, south Australia, on the different directions of the sea breeze at Stansbury and Adelaine. In the afternoon each sea breeze is the resultant of the local Gulf sea breeze and the Continental sea breeze. (After Physick & Byron-Scott, 1977.)

2.32 *Effect of mountains*

The presence of mountains within 10 or 20 kilometres from the coast may have similar effects to those described above caused by a large land mass. Due to heating on the slopes, local winds form in the mountains and may also be felt for some distance around them.

The influence of mountain forcing on the direction of wind rotation has been the subject of a theoretical study (Kusuda & Alpert, 1983) which helps to explain some of the distortions of some observed sea-breeze hodographs.

An example of the effect of mountains on local sea-breeze development can be seen in the north-east of Scotland. The map in figure 2.8 shows the presence of mountains of over 1000 m which exist 20 or 30 km inland from the coast and their influence on the direction of the sea breeze at both Kinloss in the north and at Aberdeen to the east. The mountain-effect at Kinloss rotates the breeze direction in a clockwise direction, thus adding to the rotation caused by the Earth's rotation. At Aberdeen, however, the mountains should rotate the breeze in an anticlockwise direction, and reduce the rotation caused by the Coriolis force. An effect in the same direction would also be expected from the change in direction of the coastline between the two sites.

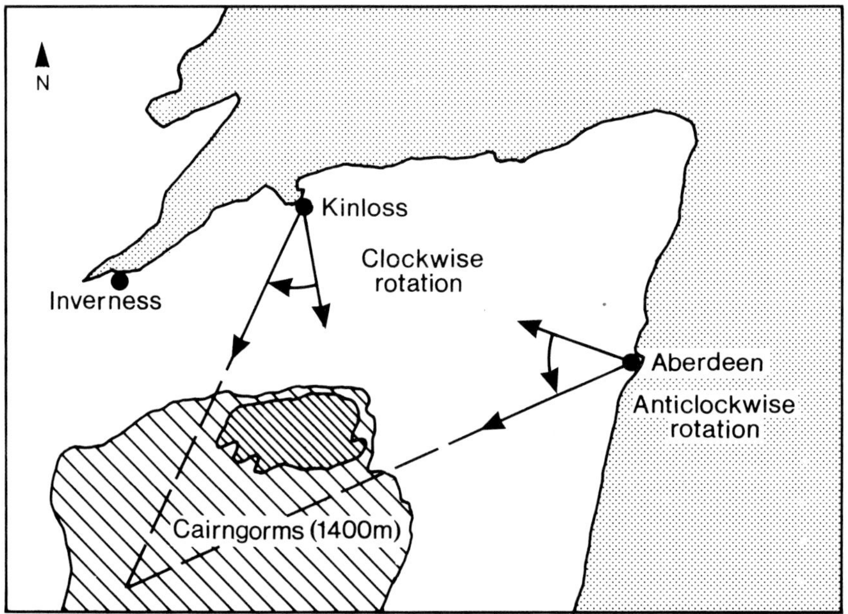

Figure 2.8. Map showing the change in direction of the sea breeze during the day at Kinloss and at Aberdeen. The change is shown towards the mountains (hatched) and the larger land mass inland. The direction at Kinloss is clockwise and that at Aberdeen is anticlockwise.

The sea breeze hodographs at Kinloss and Aberdeen are compared in figure 2.9(*a*) and (*b*) and the effects produced are seen to be very marked. The hodograph at Kinloss shows a pronounced swing in a clockwise direction, whereas that at Aberdeen shows a different pattern during the period measured.

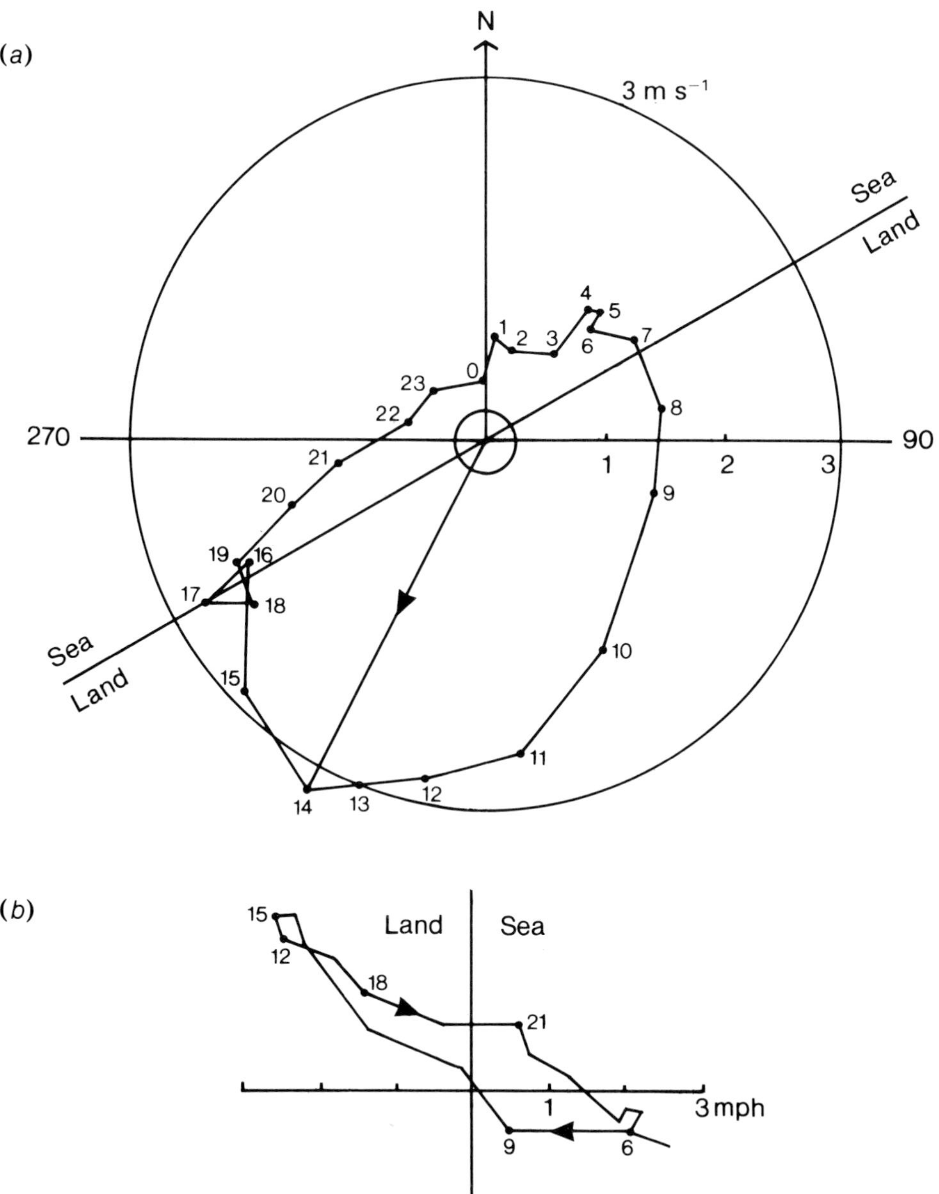

Figure 2.9. Sea-breeze hodographs compared for (*a*) Kinloss and (*b*) Aberdeen. Kinloss shows an almost steady clockwise turn, but Aberdeen shows very little rotation.

2.33 Complex coastal and mountain effects

The area of Northwest Washington, USA, shown in figure 2.10, has been the subject of much extended field work on the sea-breeze patterns (Staley, 1957) and many hodographs are available for comparison. The complex coastline and the nearby mountains are responsible for many different forms of hodograph.

Rotations of hodographs were seen in both clockwise and anticlockwise directions. For example, the hodographs of the two stations marked on the Juan de Fuca Strait rotate in opposite directions during the day. This is probably due to their different positions relative to the mountain mass to the south.

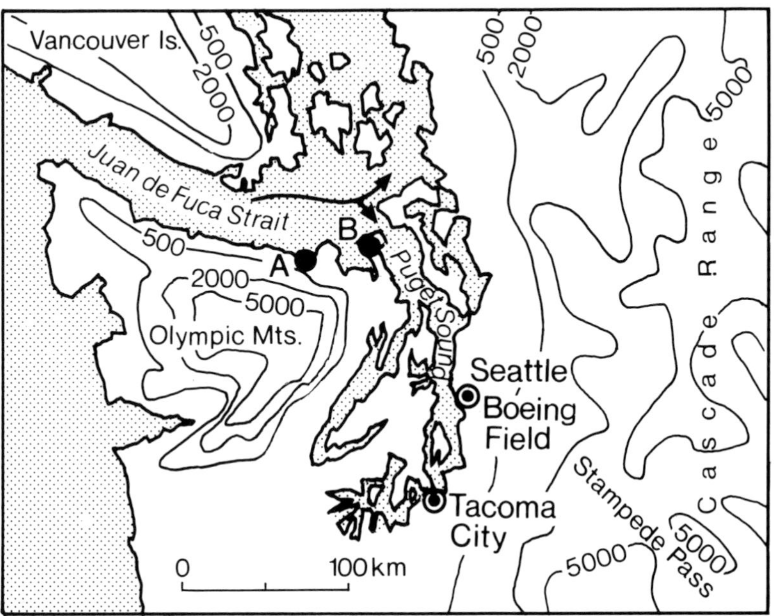

Figure 2.10. An area of Northwest Washington, USA, where much field work on wind hodographs has been carried out. The sea-breeze hodographs at the stations marked A and B rotate in opposite directions, as do those at Boeing Field and Tacoma City. (After Staley, 1957.)

Two more examples, whose hodographs are shown in figure 2.11, are Boeing Field and Tacoma City; although they are close to each other they display clockwise and anticlockwise rotation, respectively. These different rotations are thought to be due to the presence of pressure patterns which are connected with strong winds blowing from the west every day up the Stampede Pass. This study showed a range of pressure influences extending over 140 km.

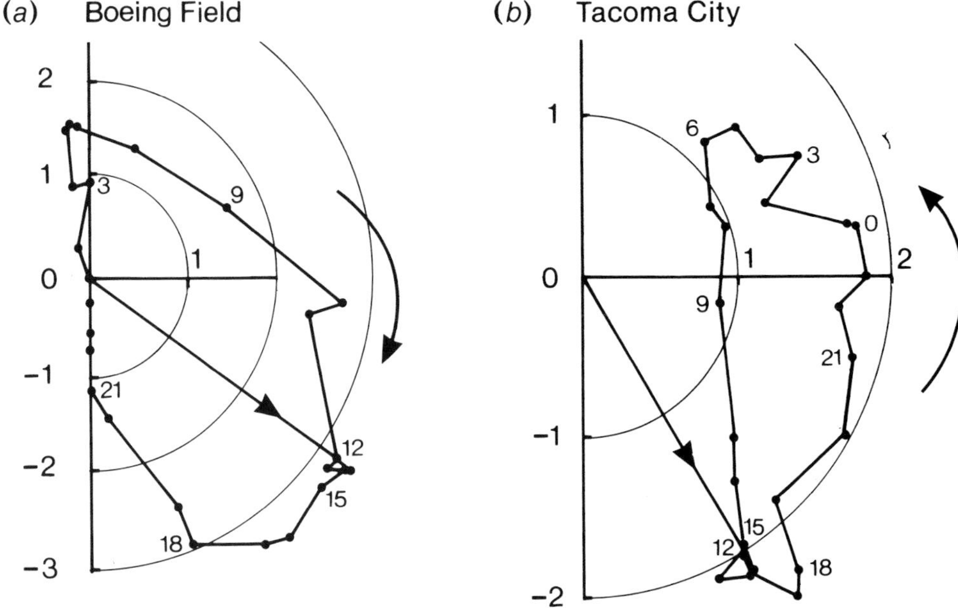

Figure 2.11. Sea-breeze hodographs for August in Northwest Washington, USA, (*a*) clockwise rotation at Boeing Field, (*b*) anticlockwise rotation at Tacoma City. (After Staley, 1957.)

2.4 Horizontal extent of the land–sea-breeze system

The extent of land and sea breezes varies considerably over different parts of the world; a summary of the literature made by Atkinson (1981) showed maximum inland penetration at different sites varying between 30 and 300 km. Many of these results were isolated measurements and may even be rare occurrences, so it is worthwhile looking further at some parts of the world where extensive series of measurements have been made.

2.41 Distance inland

AUSTRALIA

In the 1950s it was discovered that the sea breeze could be detected at what seemed to be vast distances inland. Measurements of 'inland surges connected with the sea breeze' were made at several points along the south coast. Starting from Esperance in Western Australia, using a line of stations measuring pressure, temperature and humidity, sea-breeze surges were traced as far as Kalgoorlie, 290 km from the coast (Clarke, 1955). Further east, in Victoria and New South Wales, similar inland penetration was found. Figure 2.12 shows four examples

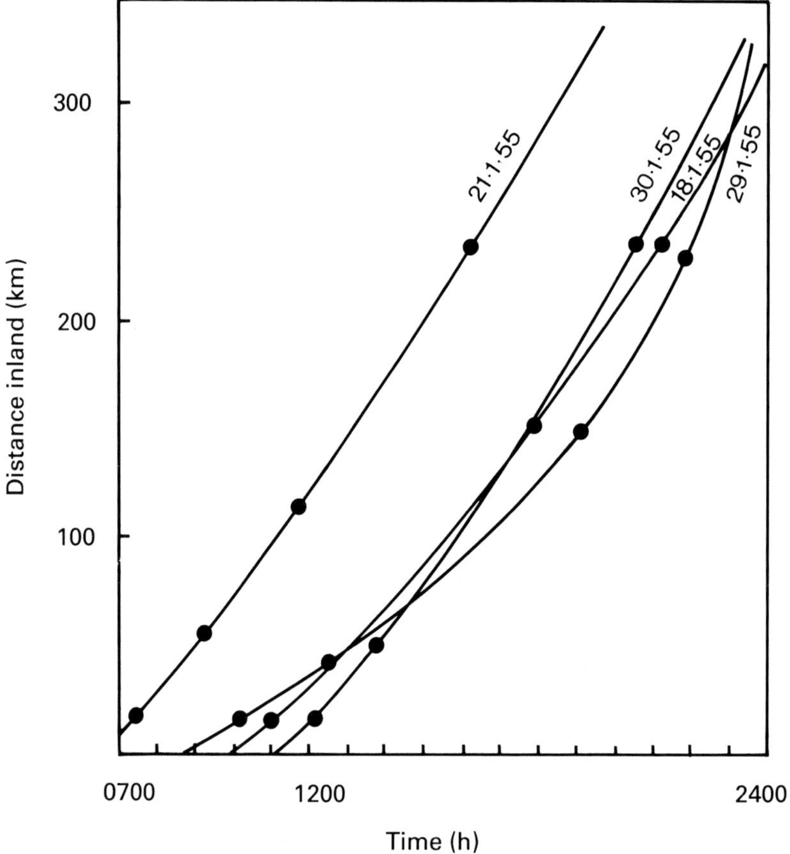

Figure 2.12. Deep inland penetration of the seabreeze. The four curves show the advance of sea-breeze surges from Esperance, near the coast, to Kalgoorlie, 290 km inland, in Western Australia. (After Clarke, 1955.)

of the progress of sea breeze from Esperance to Kalgoorlie during January 1955; it can be seen that many of these cases show an acceleration during the advance inland.

INDIA

The prevailing winds over the Peninsula during March to May have a mainly westerly component, and from the west coast the sea breeze reaches as far inland as Poona, 130 km inland. (Rao, 1955) On the east coast, at Madras, the sea breeze has been measured 30 km inland but a study further north at Jagdalbur by Banerjee, Chowdhury & Bhattacharjee (1975) established its regular presence 150 km inland at this site.

USA

There is a long history of sea-breeze observations here, both on the east and west coasts and also at the Great Lakes (lake breezes!). The spread of the sea breeze inland from the Pacific coast is complicated by the effects of the mountains, and a distance of over 100 km has been established there (Fosberg & Schroeder, 1966). From the coast of Texas, distances of 60 km have been recorded (Hsu (1970)). Around the Florida peninsula and at the coast further north, distances of 50 km were established over 100 years ago, as already described in Chapter 1.

BRITAIN

In sunny weather in the summer an inland sea-breeze advance of 30 or 40 km is not unusual from many parts of the coast. The main studies of inland penetration have been carried out over a continuous period of 12 years on the south coast (Simpson, Mansfield & Milford, 1977). Here it was found that the sea breeze occasionally reached 100 km from the coast. Isolated cases approaching this distance have been found as other places, for example at Harrogate, in Yorkshire, which is 86 km from the east coast. The observer, who was playing golf on hilly exposed ground to the north-west of Harrogate, caused some amusement among his fellow golfers when he suggested that the steady easterly wind which sprang up at 1845 GMT might be the sea breeze! (its authenticity was established by examining other meteorological records for the day (Fergusson, 1971).)

2.42 Distance out to sea

Few measurements exist of the distance to which the land- and sea-breeze system extends out to sea. A very striking indication of the width of the sea breeze can be obtained from one of the Gemini XI photographs of 1966. This photograph, shown in figure 2.13, looking roughly north, shows the southern tip of India, with Sri Lanka visible to the right. Sea-breeze clouds can be detected along both coasts, but the clearest feature is the cloud-free zone extending along the west coast. The most likely cause of this clear space is the descent of air in the sea-breeze circulation dispersing all cloud. The width of this zone, measured from the photograph, is between 100 and 120 km.

The weather map of 1200 GCT for this day, 5 hours after the photo was taken, shows winds blowing towards the shore along all coasts. Air over central India was calm and a slight low pressure system was over the northern part of the sub-continent. Temperatures of coastal air were about 27.5 °C and 4 to 5 °C warmer inland, conditions typical of a sea-breeze day. So the seaward extent of the sea breeze was precisely measured, perhaps for the first time (Stevenson,

Figure 2.13. A cloud-free zone is seen extending 100 km over the sea along the western coast of South India. This zone is produced by the descending air of the sea-breeze circulation and indicates its scale. Photo taken by Gemini XI, on 14 September 1966.

1969). Distances of the extent of the sea breeze similar to this were deduced from aerial measurements made further north in the Bombay area (Dixit & Nicholson, 1964).

Other extensive measurements of sea-ward spread have been made in the Baltic Sea and the Gulf of Bothnia. Along the east coast of Sweden it was found (Grenander, 1912) that the land- and sea-ward extents were about the same, a distance of 45 km. Further north, in the Gulf of Bothnia on the Finnish coast, the distances were once again equal but only extended 20 km (Rossi, 1957).

2.43 Frequency of sea-breeze penetration

The average number of days per month when the sea breeze reached Renmark, 290 km from the West Australian coast, is shown in figure 2.14. This is the result of continuous measurements over six years. As would be expected, arrivals are most frequent in the summer, but events can occur during any month of the year.

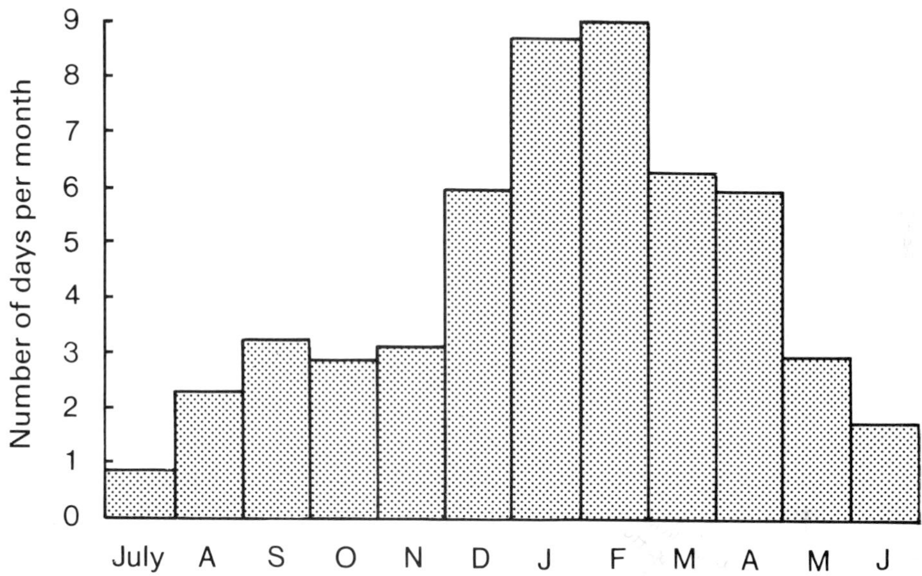

Figure 2.14. Average number of days per month when the sea breeze reached Renmark, 290 km from the West Australian coast. (Records measured over 6 years, Clarke, 1955.)

A different type of plot to show how often the sea breeze reached inland sites is shown in figure 2.15, in which the results are averaged over 12 years in south-east England. It shows for example that one sea breeze a year may be expected at Harwell, 80 km inland, but arrival at Oxford, over 100 km inland, is rarer still. The logarithm of the frequency has been plotted against distance inland and it is found that the points fall very close to a straight line.

The frequency of distance of spread of the sea breeze on the Baltic Coast, both inland and also out to sea, is shown in figure 2.16. It seems that on half of the days when a sea breeze was present it could be expected to spread about 30 km, both inland and out to sea.

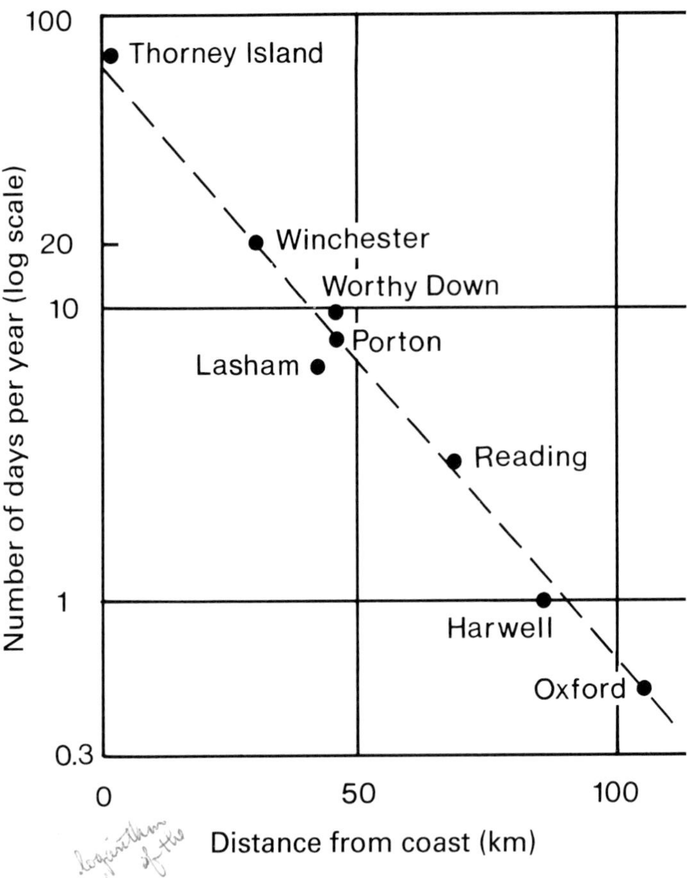

Figure 2.15. The frequency of penetration of sea breeze from the south coast of England to various points inland. Figures are given in days per year, averaged over a period of 12 years.

2.44 Horizontal extent of land breeze

Measurements of the land breeze as it extends out over the sea have been made at Wallops Island in north-east United States (Meyer, 1971). Using 100 m wind tower measurements, rawinsondes and a powerful 10 cm radar, it was established that the land breeze was a layer of cold air less than 100 m deep and reaching 25 km from the shore.

The progress of the land breeze in the winter over Lake Michigan is important because its boundary is often associated with the formation of cloud bands and Great Lake snow storms (Passarelli & Braham, 1981). The formation and persistence of a line of cloud parallel to the lake's axis gives a distance of the land breeze from the coast of 60 km.

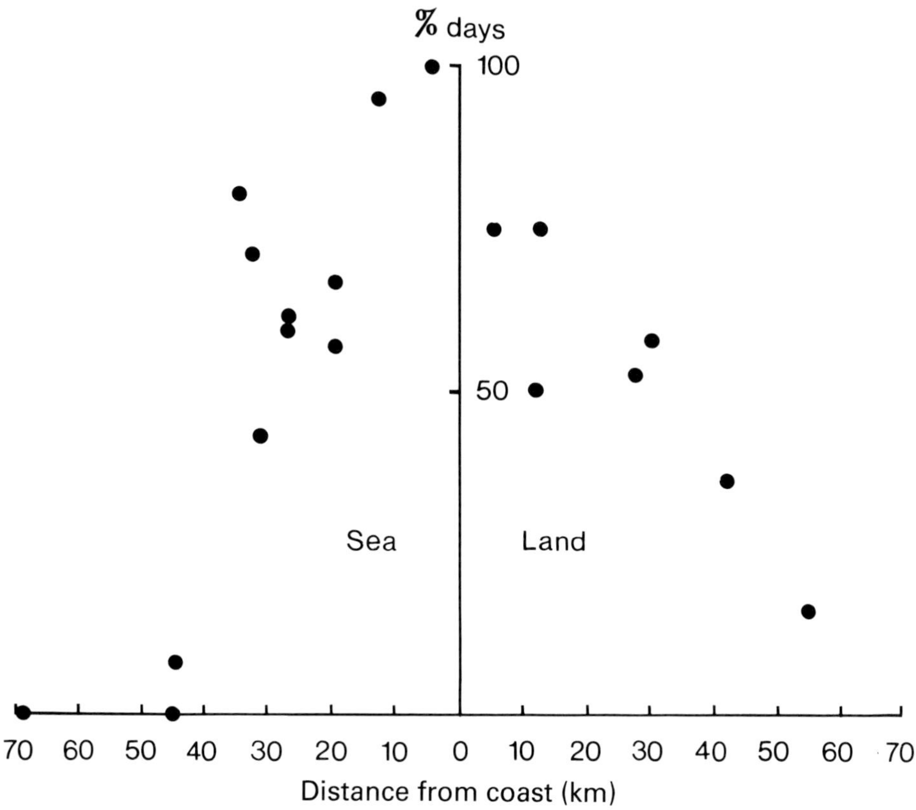

Figure 2.16. Spread of the sea breeze from the Baltic coast, both inland and out to sea. Results are given as a percentage of days when a sea breeze occurred. (After Rossi, 1957.)

A band of precipitation extending along the English Channel on 19 October 1990 was thought to have been enhanced by land-breeze circulations (Waters, 1990). This extended a distance of 60 km from the south coast on England, the same distance as measured for the land breeze in Lake Michigan.

3

Sea-breeze fronts

The onset of the sea breeze is sometimes in the form of a sudden squall, resembling a minor cold front, which is called the sea-breeze front.

This front is of special interest out at sea to yachtsmen who may expect sudden wind changes in this area. Glider pilots, many kilometers inland from the coast, can sometimes use the rising air currents formed at the leading edge of a sea-breeze front. Its dynamics are of interest to environmental scientists since the presence of the front has important effects on the distribution of airborne pollution.

3.1 Structure of a sea-breeze front

The form of the leading edge of the sea breeze was put on a firm basis by the field work of H. Koschmieder (1936) at Danzig in the 1930s. He made extensive measurements on the ground and with pilot balloons.

He established that even on calm days when the sea-breeze started early with no sudden changes at its boundary, after midday it appeared as a squall line; this was the norm along the Baltic coast on days with a steady offshore gradient wind.

The wind behind this front was found to be greater than the speed of advance of the front; the sea breeze ascended at the front and his measurements showed that it returned relative to the leading edge of the front, so he distinguished between the flow of the mixed 'sea-air' and the wind of the 'sea-breeze proper'. His measurements showed that the depth of the sea breeze was usually about 350 m and the mixed sea-air extended to about 700 m.

Extensive field measurements have been made since then of the characteristics of sea-breeze fronts, especially in USA, UK and Australia. Close similarities have been found between these fronts and thunderstorm outflows. The latter have been closely studied owing to the danger they cause to aviation. From laboratory measurements and numerical simulations of the thunderstorm outflows (Simpson, 1987) the structure of this type of atmospheric gravity current is well established.

A gravity current (or density current) is primarily horizontal and may be generated between any two fluids with a density difference. The temperature difference between the sea-breeze air and the air further inland is usually less than 3 °C, which corresponds to a density difference of about 1%.

A model of a sea-breeze gravity current front is illustrated in figure 3.1. The opposing wind from the right meets the wedge of cool dense air from the sea and forces it backwards and upwards, behind a sharp leading edge, forming a raised 'head' about twice the depth of the following steady flow. A turbulent wake lies above at the rear of this head.

Although the head may be as deep as 700 m, aircraft measurements through the sharp leading edge of the front sometimes give a transition layer between the two air masses of only a few tens of metres thick.

As the cold dense flow arrives at the leading edge it is forced upwards (as shown in figure 3.1), maintaining the sharp interface between the two flows. The upper fluid has greater velocity, but the fluid beneath has greater density: these are the conditions in which the instability named after Kelvin and Helmholtz can occur. This is manifested in the formation and rolling up of billows on the interface between two layers of different density which are moving relative to each other; it has been observed in other situations in the atmosphere. These Kelvin–Helmholtz billows grow until they become unstable and break down, thus forming the 'turbulent wake'. The vertical thickness of

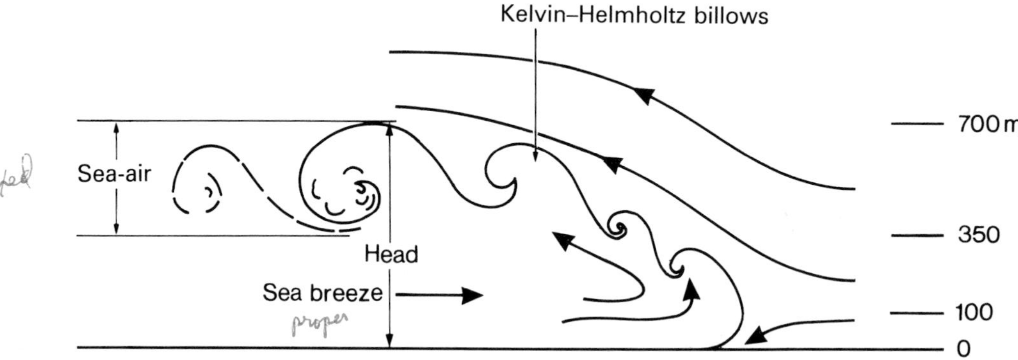

Figure 3.1. Model of a sea-breeze gravity current front. The opposing air from the right forces the dense sea breeze upwards and backwards from the front, forming a raised 'head'. Kelvin–Helmholtz billows, which form along the interface between the two fluids, grow and eventually become unstable and break down. This forms a mixed layer of 'sea-air' at about the same height as the sea breeze itself.

this breakdown layer is roughly equal to that of the unmixed sea-air approaching the front beneath it. Thus the size of the head is determined by the breakdown size of the billows and its height is found to be about twice the depth of the advancing dense sea-breeze air.

This simple model covers most of the observed features of sea-breeze fronts. The behaviour of the flow where the billows are forming is often made visible by distinctive 'curtain clouds' at the interface. These are illustrated in section 3.3.

3.11 Lobes and clefts

When the front is moving forward over the ground, another important factor has to be considered. The combined influences of ground friction and inertia form an overhanging nose at the leading edge. Up to a height of about 100 m some of the less dense air is overrun by the dense sea breeze. This leads to another form of instability, in which the leading edge of the front is divided into a series of 'lobes and clefts'. Figure 3.2 shows how most of the overrun air is diverted into clefts inside which it eventually rises through the denser sea-breeze air. The spacing of these clefts is roughly 1 km.

The presence of lobes and clefts means that the flow can never be strictly

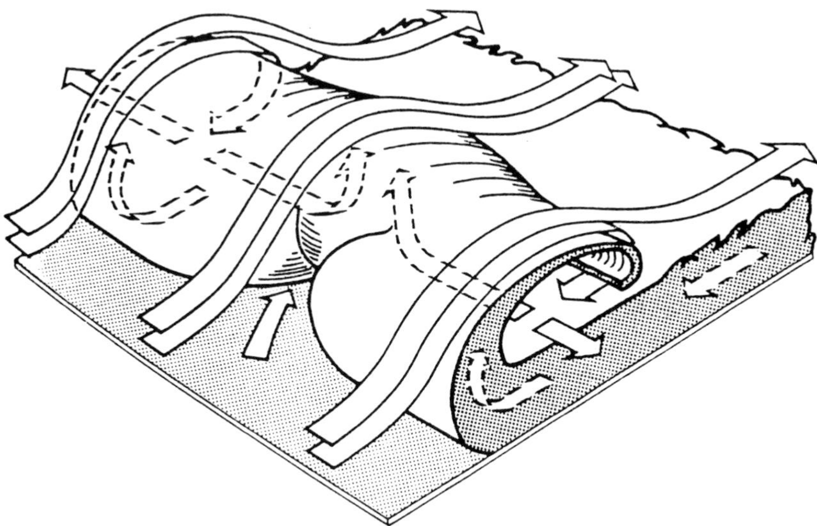

Figure 3.2. Schematic diagram of the flow at a sea-breeze front, in which the leading edge is divided into a series of lobes and clefts. Some of the warmer air is overrun and is ingested in the cleft in the centre.

two-dimensional, as suggested in figure 3.1, over any great lateral extent. Any particular set of Kelvin–Helmholtz billows will not extend more than about a kilometre along the line of the front.

3.12 Measured profiles of the front

Cross-sections of sea-breeze fronts have been measured by successive traverses using instrumented aircraft. One such cross-section is shown in figure 3.3. This profile was built up from six transverse flights at different heights using an instrument which gave high-speed-response humidity measurements. The humidity profile is plotted here as moisture mixing ratios, in grams of water per kilogram of air. This shows clearly the unmixed sea air beneath 400 m, with mixing taking place at higher levels, where a large area of dry air has been engulfed.

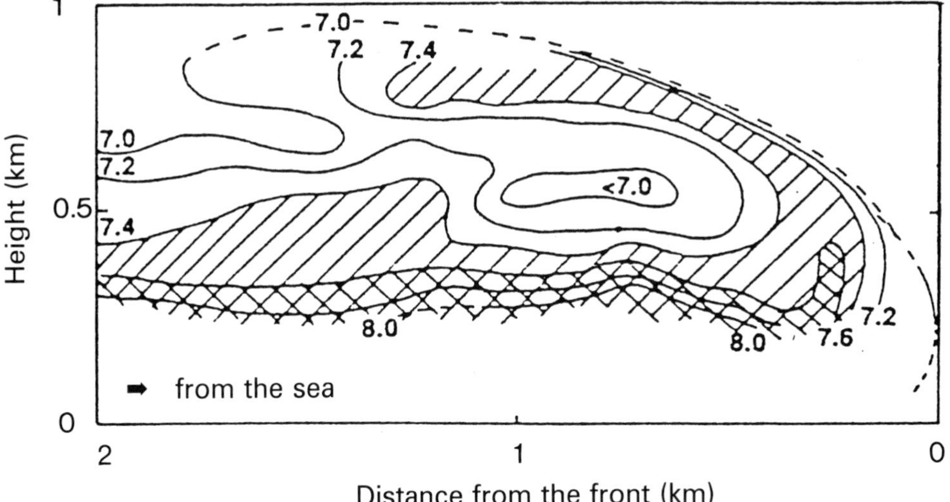

Figure 3.3. Cross-section of a sea-breeze front from traverses of an instrumented aircraft, showing humidity mixing ratios (in g kg^{-1}). Sea-air is unfixed up to about 400 m; mixing takes place at higher levels with a large area of engulfed dry air. The hatching and shading show the boundaries of the moister air of 7.4 and 7.6 g kg^{-1} respectively.

The front shown in figure 3.3 was measured in almost calm air, but the profile of a sea-breeze front can be greatly modified by any opposing wind. This lengthens the head profile and reduces the height of the foremost point, or nose. If the strength of the opposing wind continues to increase with time it may become strong enough to bring the front to rest; the foremost point is then on the ground and the front forms an arrested wedge of roughly uniform slope.

Laboratory experiments, confirmed by some full-scale atmospheric measurements (Simpson & Britter, 1980), have shown that the effect of a head wind is to flatten the head and to reduce the speed of advance of the front by about three-fifths of the applied wind. Figure 3.4 shows dense salt solutions moving in a laboratory tank of fresh water. In (*a*) the surroundings are calm, but in (*b*) there is an opposing flow of fresh water. It can be seen that the opposing flow acts to reduce the slope of the front.

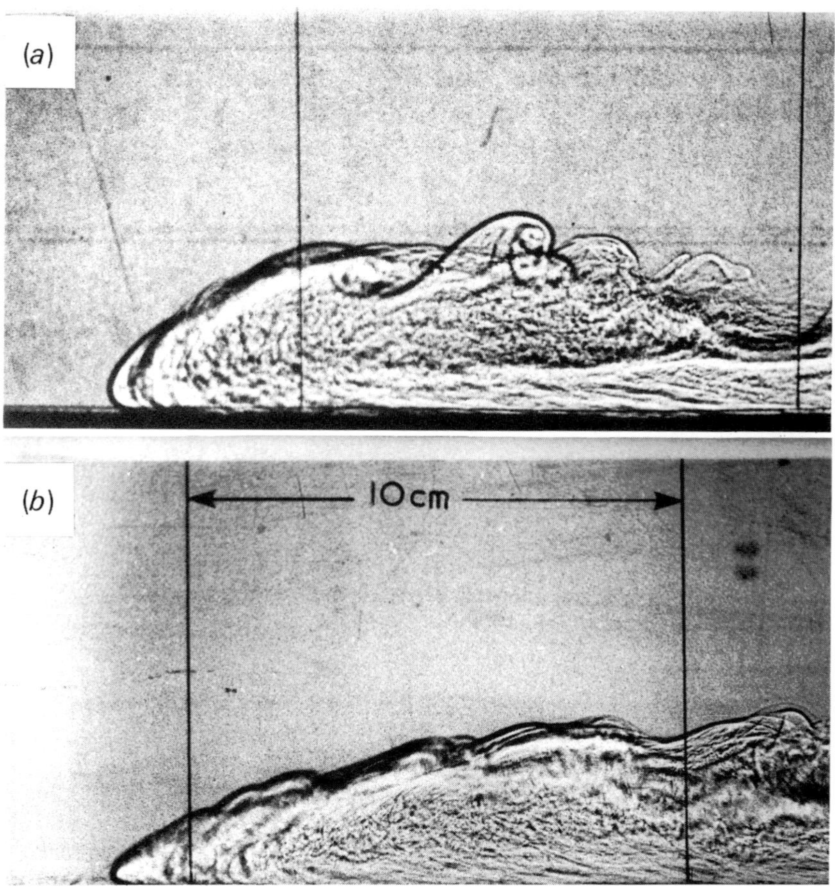

Figure 3.4. Gravity current fronts in the laboratory. (*a*) Front in calm surroundings. (*b*) Front affected by head wind, showing the flattened head.

3.2 Generation of sea-breeze fronts (frontogenesis)

The generation of a front between two air masses of different density depends mainly on the strength of the converging winds.

If the offshore wind is strong enough, a fully developed sea-breeze front can be generated near the coast. This may remain stationary or move slowly inland while the sea breeze itself continues to blow behind the front and air ascends along the line where the converging winds meet.

In calm weather or with only light opposing wind, the sea breeze is able to extend its influence gradually both inland and out to sea. However, as the sea-breeze air spreads inland its properties are modified by travelling over the heated ground; for example, on a calm day the temperature of the sea-breeze air may closely approach that of the land air at a distance of about 10 km from the coast.

In this area where a horizontal density gradient exists, gravitational forces might be expected to generate a front, but in the presence of strong thermal convection gravitationally driven horizontal flows are weak, the air is mixed and frontogenesis does not occur.

Laboratory experiments on the effects of turbulence on fronts in gravity currents which confirm this sequence are described in detail in Chapter 11.

In the late afternoon or evening vertical mixing will eventually decrease sufficiently for the horizontal temperature gradient in the cool dense sea-air to tighten up and create a sharp gravity current front of dense sea-air which then accelerates and continues to move inland (Reible, Simpson & Linden, 1993).

Measurements of such a temperature gradient in figure 3.5 show how it developed into a sharp front when the turbulent mixing decreased late during the day. Potential temperature is plotted, because the density of the air is inversely proportional to its potential temperature, which is the temperature the air would attain if it was transferred adiabatically to ground-level pressure. The information in figure 3.5 was measured from an instrumented aircraft making several horizontal passes through the boundary of the sea-breeze air, on a June day in southern England. In the first flight the average horizontal temperature gradient in the sea-breeze air was 0.28 °C km^{-1}, reducing to a very small gradient in the land-air. In the second flight, made two hours later in the evening, the temperature gradients were both close to zero, and a sea-breeze front had formed with a sharp frontal change in temperature through a distance of about 100 m.

In conclusion, given sufficient contrast between land and sea temperature, the formation of a sea-breeze front depends on the balance between converging winds, which act to form a front, and the vertical mixing over the land, caused by thermal convection, which acts to prevent its formation.

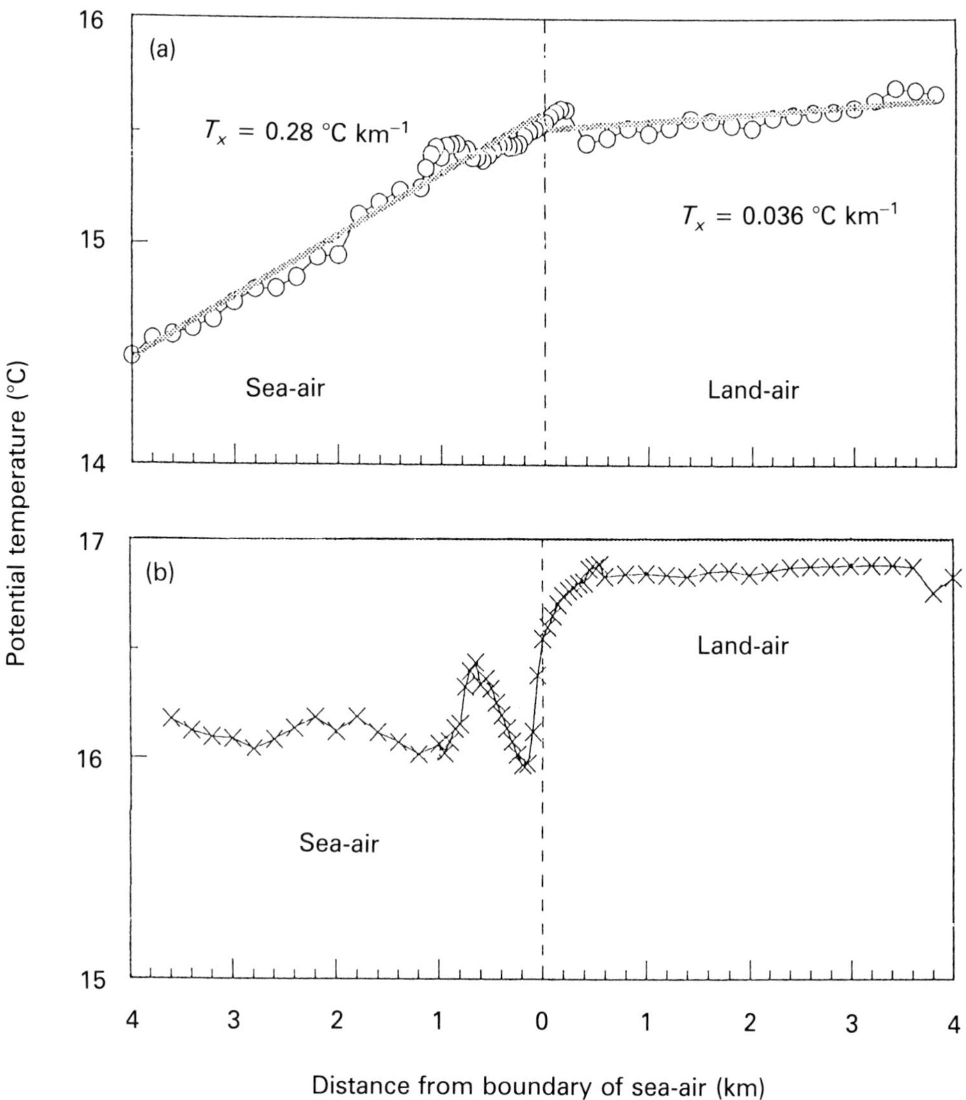

Figure 3.5. Frontogenesis on a calm day: Horizontal temperature profiles are measured at a height of 300 m, through the boundary of the sea-breeze air. Average temperature profiles (*a*) in the early afternoon, 1505–1605 GMT. T_x is the horizontal temperature gradient. (*b*) Two hours later, 1730–1805 GMT, showing a sharp sea-breeze front.

It is even possible for a sea-breeze front to form on a day of light onshore wind, provided that the land–sea temperature difference is large enough. On such days, the convergence of the winds will be small and a front will not form until late in the day, 50 km or more from the coast.

3.3 Clouds at the sea-breeze front

The existence of a sea-breeze front can often be deduced from the appearance of distinctive forms of cloud. The air which is forced to ascend in the convergence zone of the sea-breeze front may reach condensation level and on a clear day this may cause a line of clouds to form in an otherwise cloudless sky. On a day when the sky is full of small fair-weather cumulus, the front can make its presence visible by greater cloud development along a line. There may be no more cloud on the seaward side of this line. An example of both these effects from near the south coast of England is shown in figure 1.4.

A characteristic way in which the sea-breeze front can make itself apparent is in the formation of fragments of cloud in the moist sea-air, which has a much lower condensation level than that of the cumulus clouds inland. Ragged veils or curtains of a very distinctive form of cloud can often be seen rising rapidly, as shown in figure 3.6. This is a view of a sea-breeze front seen from a glider, soaring close to the front, just below the base of the cumulus clouds. Beneath us to the left is the rising turbulent moist sea-air, which is beginning to condense into ragged fragments of cloud along the slope of the front.

Figure 3.6. Ragged veils or curtains of cloud outlining the front of the sea breeze, which is blowing from the left. Seen from a glider soaring in the clear air on the landward side.

Figure 3.7. View of sea-breeze front from above. This front is clearly marked with haze and shows low cloud forming on two advancing lobes of the sea breeze, which is advancing from the right. (Photo by H. Howitt.)

Viewed in the air from the side these clouds can be seen to be sloping, as shown in figure 3.7. This front was clearly marked by smoke haze, and low cloud can be seen forming on two advancing lobes.

Seen from the ground, ahead of the front, the slope cannot usually be distinguished and the ragged cloud merely appears to be hanging below the base of the cumulus clouds. In the example of figure 3.8 the curtain clouds show that the condensation level of the air in the sea breeze forms a sharp cloud-base 300 m lower than that of the inland cumulus. At the front air rises with less mixing, and therefore less dilution, than in thermal convection. So at the front the lowest cloud fragments are lower than the clouds inland, even if the humidity is the same at the front.

Industrial smoke and haze sometimes outline the advancing boundary of the sea breeze and can make the outline of the front clearly visible as it approaches. This has already been illustrated in Chapter 1 and examples in the industrial north of England and also in California are described in Chapter 6, which deals with airborne pollution.

Figure 3.8. Curtain clouds, with lower base, seen from the ground at the sea-breeze front near Winchester, 8 August 1962.

3.31 Land-breeze fronts

The clouds which form at land-breeze fronts have only rarely been photographed, so the picture in figure 3.9 is of especial interest. This shows a land-breeze front moving off the north-west coast of Hokkaido, Japan. Clouds form only at places where the moist sea-air mixes with the cold land-air; they can be seen in the billows above the cold flow (moving from the right) and also appear in fragments near the ground in the overrun moist air (Tsuboki, Fujiyoshi & Wakakama, 1989).

3.4 Advance of sea-breeze front

The time of onset of the sea breeze has been recorded all over the world at many coastal sites, but its subsequent advance inland has been much less fully studied. The spread inland of the sea breeze is of special interest to those concerned with air quality, especially with airborne pollution; as discussed in Chap-

Figure 3.9. Land-breeze front off the north-west coast of Hokkaido, Japan. Clouds form where the moist sea-air mixes with the cold land-air moving from the right. Taken at 0800 h on 6 February 1992 by Y. Fujiyoshi of Nagoya University.

ter 2 many individual case studies exist when the sea breeze front travelled as far as 100 km from the coast.

3.41 Methods of observation

Some of the earliest measurements of the inland invasion of the sea breeze were made in the nineteenth century in Massachusetts, USA, by Davis, Schultz & Ward (1890). A number of observers, spaced out along a line at right angles to the coast, were asked to note the time of arrival of the wind from the direction of the sea and the data were collected by riding along their line on a horse. Since the invention of the telephone and the automobile this kind of work has been much speeded up, but the greatest advance has been in the use of various types of 'self-recording' instruments. The most obvious of these is an anemometer, recording the direction and strength of the wind, but such instruments are expensive. A very good indication of the arrival of the sea breeze can be obtained by much simpler instruments measuring both temperature and humidity; this enables the change in dewpoint in the sea-air to be measured. Dewpoint is a very good indicator of a sea breeze, as this property of the air is not changed by heating or cooling.

During the 1970s and later numerous weather station packages have been marketed which record all required data electronically in a form which can be re-examined at a later date.

The simple instrumental records shown in figure 3.10 illustrate most of the features seen in the sea-breeze passage inland from the south coast of England on one day in June 1963. The wind strength and direction records were measured by a Dynes anemometer, which originated in the last century. The relative humidity was measured by a hair hygrometer and the temperature by a bi-metallic coil. For a long period of time these three instruments were in standard use by the Meteorological Office.

The first station, Thorney Island, lies close to the coast. It can be seen that the first onset of the sea breeze was soon after 1000 GMT, but although moist

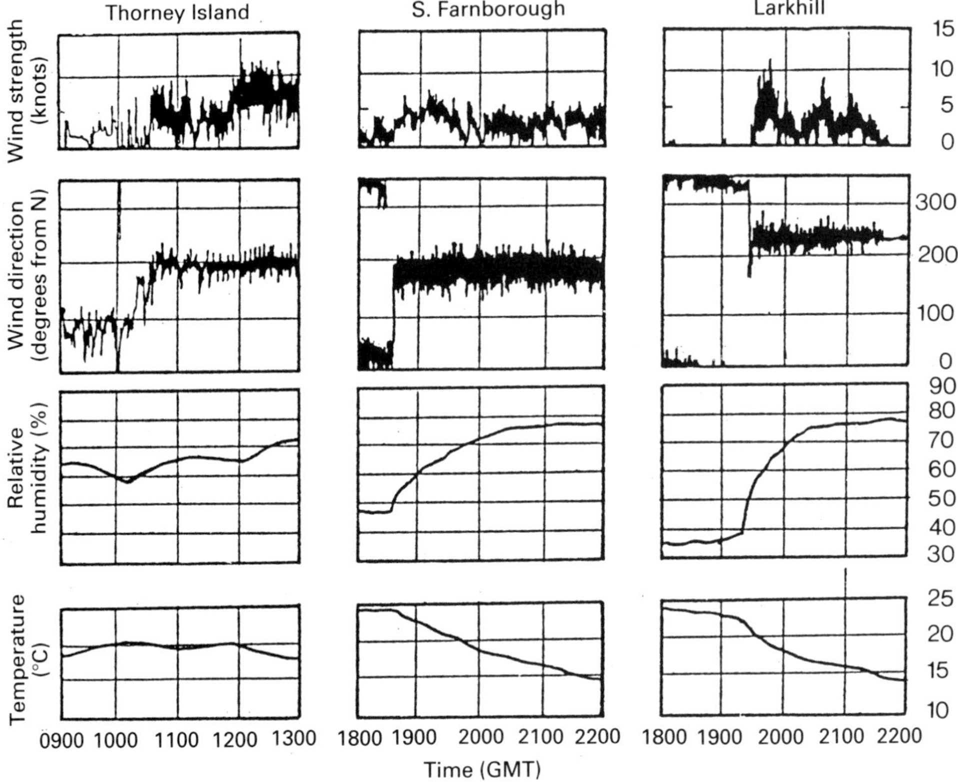

Figure 3.10. Simple instrumental records of the changes in wind, humidity and temperature during the passage of a sea-breeze front on a summer day (11 June 1963) in southern England. At Thorney Island, near the coast, the onset of the sea breeze is gradual. The other two sites, at 30 km and 40 km inland from the coast, show sudden frontal onset of the sea breeze later in the day.

air had started to blow, it was not until 30 minutes later that the sea breeze was fully established. After 1040 GMT the sea breeze blew steady at 5 knots (2.5 m s^{-1}) from a direction of 200°.

The second station, S. Farnborough, is north of Thorney Island, but 30 km from the coast. At 1830 GMT the sea breeze was now clearly in the form of a front, with an instantaneous change in wind strength and direction. The wind strength was a little less than it was at the coast, but the direction was identical. There was a sudden shift in both humidity and temperature. After the passage of the front the temperature continued to fall, as the time was near sunset. The relative humidity continued to rise, but the calculated dewpoint, a conservative property of the air mass, which increased suddenly at the arrival of the front, remained almost constant.

The third station, Larkhill, is north-west of Thorney Island, and at a distance of 40 km. The jumps in relative humidity were even more marked, and there were two new features visible in the wind records. Firstly, the wind had veered in direction and was now blowing from 240°. Secondly, there was a regular oscillation which can be seen in the wind strength, although the direction remained constant. This indicates the formation of the bore-wave which forms at the sea-breeze front when it is still actively moving inland at sunset.

3.42 Frontogenesis during the the advance

The records examined above, made in June 1963, gave hints of the gradual development (or frontogenesis) of the sea-breeze front from an initial state near the coast where the establishment of the sea breeze was spread over several kilometres and the gradients of temperature and humidity were still both very weak.

The development of the front over the same locality was again investigated on a very similar day ten years later, but this time with the aid of many more ground measurements, and backed up with seven pilot balloon ascents and two flights in an instrumented light aeroplane (Simpson, Mansfield & Milford, 1977). The results are shown, in figure 3.11, which gives the times of onset of the sea breeze at various points inland.

In the early stages and until about 1500 GMT the onset of the sea breeze was quite gentle and it was not always possible to determine very accurately the exact time from the recording instruments. A smooth curve has been drawn through the points, but onset was diffuse until about 1400 GMT when the advance seemed almost to have ceased. By 1600, however, a front had started to advance and accelerated inland during the next three hours.

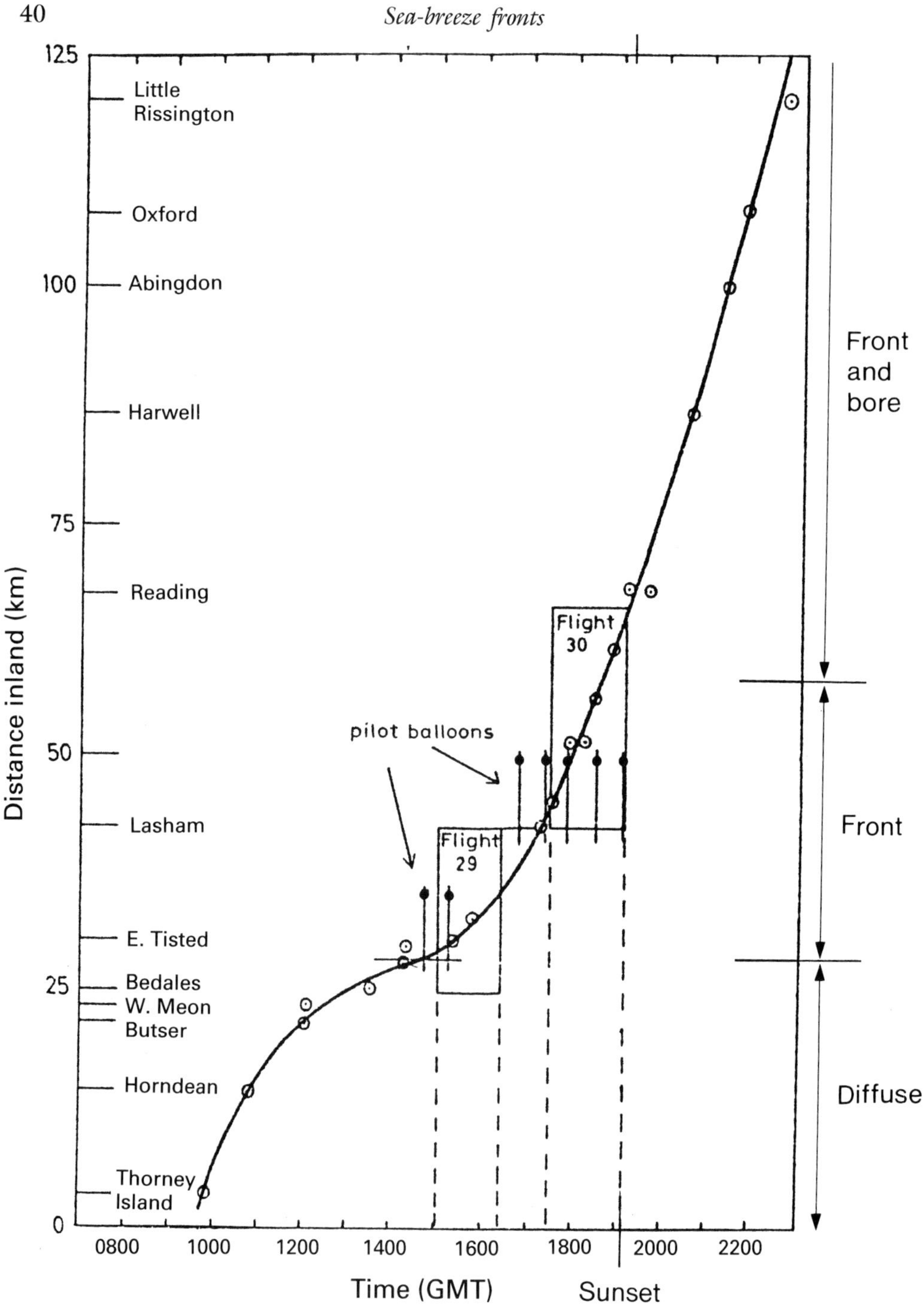

Figure 3.11. The time of onset of the sea-breeze in southern England during 14 June 1973, close to line 1 °W. In the early stages until 1500 GMT the onset was gentle. By 1600 GMT a front had formed, which accelerated inland.

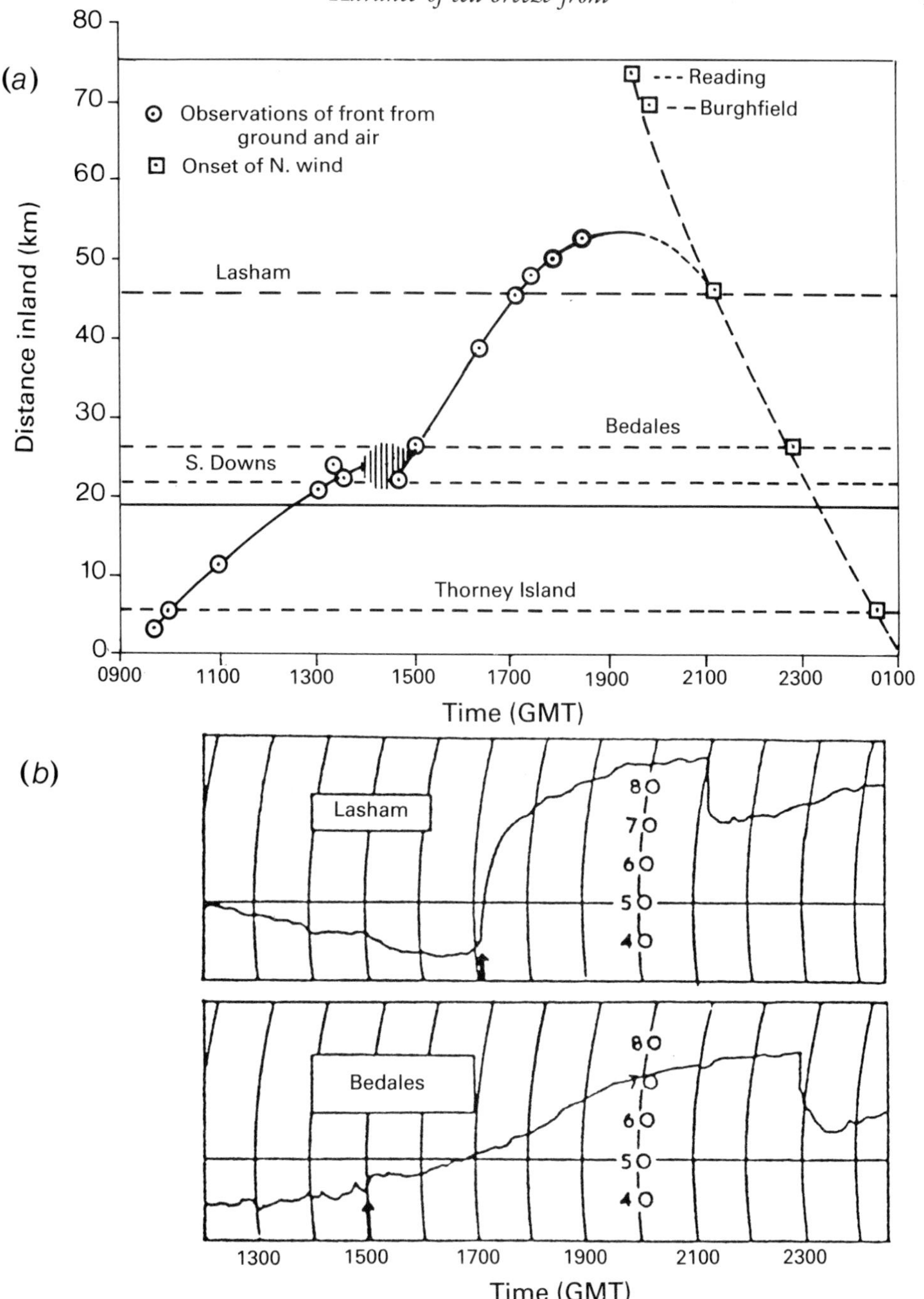

Figure 3.12. Advance and retreat of sea-breeze front in southern England on 1 June 1966. (*a*) Position of the front during the day, showing a diffuse period at about 1500 GMT before formation of the front. Later, the front moves back towards the coast. (*b*) Relative humidity (%) at two stations marked above, showing the sharp return of dry air as the front receded. Arrows show the time of onset of the sea breeze.

3.5 Retreating sea-breeze fronts

During the study of sea-breeze fronts in southern England, which lasted ten years, fronts could sometimes be seen south of Lasham (45 km from the coast) moving southwards in the late evening. The fronts which passed Lasham usually travelled a further 20 km, or even 50 km. However, in a few cases well-developed fronts returned towards the sea even after travelling 50 km inland.

One example was on 1st June 1966 (Simpson, 1974). There were light winds in the early part of the day. When the sea breeze reached the South Downs, 25 km inland, at about 1400 GMT there was not yet a sharp front, but later in the day a strong front began to accelerate inland and passed Lasham at 1700 GMT. Soon after this a line of increasing north wind was observed moving steadily south. Figure 3.12(*a*) shows how the progress of the sea-breeze front was affected as this northerly wind forced it back towards the coast. Along the line of observation it moved at a steady 3.5 m s^{-1}, and the humidity traces in figure 3.12 (*b*) show that a sharp boundary still existed during this retreat.

3.51 Sensitivity to opposing winds

The formation of sea-breeze fronts and their rate of advance are both very sensitive to the strength of any opposing wind.

An examination of three successive days in southern England (21,22 and 23 August 1972), illustrates this sensitivity (see figure 3.13). During the first two days an almost stationary anticyclone over Ireland maintained north winds over southern England, on the third day the anticyclone drifted south and the winds became light.

DAY 1 (21 AUG.)

With offshore winds of 7 m s^{-1} no sea air reached the land until 1500 GMT when the sea breeze began to blow at Southsea, near Portsmouth. The humidity trace at Southsea showed a series of six oscillations as the sea-breeze front passed backwards and forward across the recording station. Eventually, at 1800 h the

Figure 3.13. Sensitivity of sea-breeze behaviour to small wind differences on three successive days in southern England as it advanced along a line 1 °W. (*a*) Distance penetrated inland by the sea breeze on the three days in 1972. On 21 Aug. the wind was strong enough to bring the front to rest. On 22 Aug. a change of wind forced the front to retire. On 23 Aug. the wind was light and the sea breeze moved at least 75 km inland. (*b*) Relative humidity and temperature at Southsea on 21 Aug. (*c*) Relative humidity at four stations on 22 Aug.

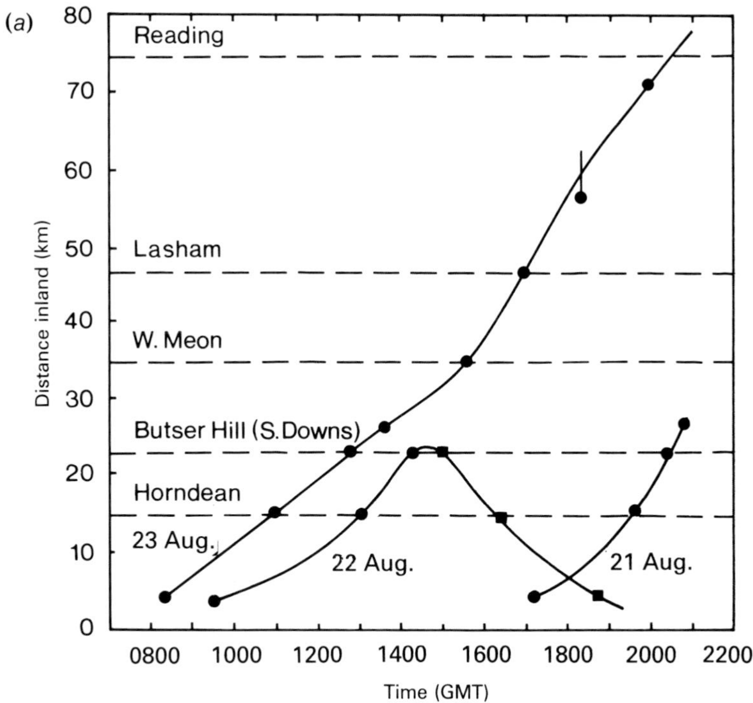
(a)
Distance inland (km)
Reading
Lasham
W. Meon
Butser Hill (S. Downs)
Horndean
23 Aug.
22 Aug.
21 Aug.
0 10 20 30 40 50 60 70 80
0800 1000 1200 1400 1600 1800 2000 2200
Time (GMT)

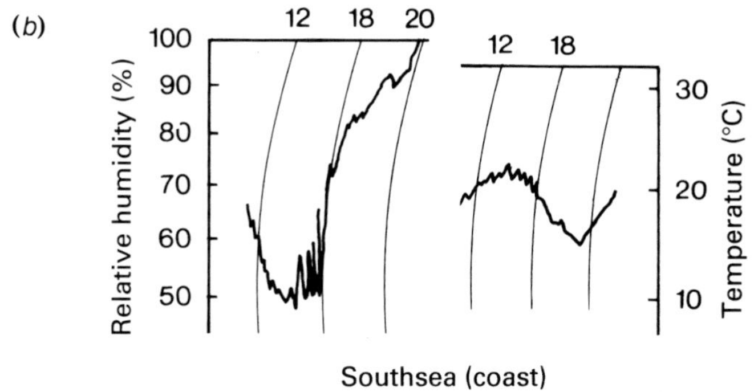
(b)
Relative humidity (%)
100 90 80 70 60 50
12 18 20
12 18
30 20 10
Temperature (°C)
Southsea (coast)

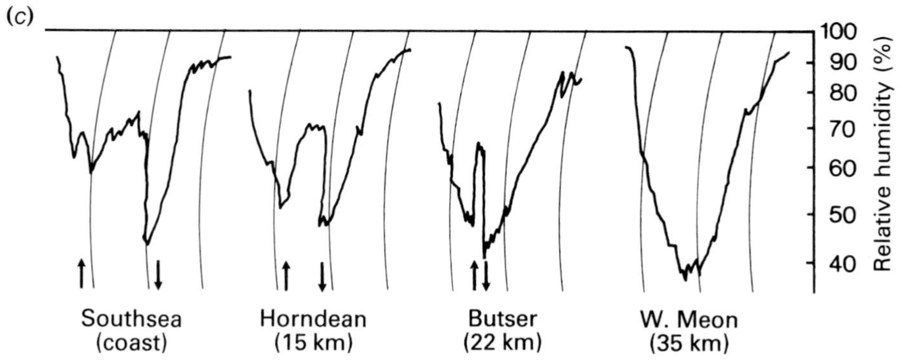
(c)
100 90 80 70 60 50 40
Relative humidity (%)
Southsea (coast)
Horndean (15 km)
Butser (22 km)
W. Meon (35 km)

sea breeze was established and began to move inland. It had reached 15 km inland by sunset and was last traced two hours later, 25 km from the coast.

DAY 2 (22 AUG.)

On this day the sea breeze moved inland at Southsea at 1000 GMT and advanced 22 km just beyond Butser on the South Downs. The humidity traces for this day show 7 hours sea breeze at Southsea, 3 hours at Horndean, 15 km inland, and less than one hour at Butser. These stations show clearly the limits of the time when the sea breeze was blowing, with the sharp return to land-air properties when the front retreated. The stations further inland recorded no sea breeze. The sharp front retreated southward, recrossing the coast at about 1900 GMT.

DAY 3 (23 AUG.)

With the light wind of this day the sea breeze passed inland before 0900 GMT. In the afternoon a well-developed front was established, travelling at a steady speed of 8 km h^{-1}. It passed Lasham, 45 km inland, at 1700 GMT and continued to advance well after sunset.

3.52 Lateral distortion of a front

The progress inland on 5 June 1973 of a 50 km section of a sea-breeze front from the south coast of England was investigated with the aid of an instrumented motor-glider, a high-power radar, and ground-based autographic records. As figure 3.14 shows, in the western section steady progress was made, but further east the front retreated back to the coast after reaching 20 km inland. During the afternoon pressure rose over the south-east corner of England relative to the rest of the country, causing a strong pressure gradient orientated north-east–south-west to develop over the extreme south-east. This halted the advance of the sea-breeze front inland in this area and later caused it to retreat south-westwards. This effect would be present whenever the gradient wind is east or north-east but the rarity of its occurrence on the 1 ° line is explained by its distance from the east coast. Under these conditions the nearest coast is the Thames Estuary, 150 km to the east-north-east. With the wind speed of 7 m s^{-1} which existed on 5th June in the extreme south-east, the influence of the east coast would take 6 hours to reach this line and only become important a few hours after this.

3.53 Sea breeze retreating within 20 km of the coast

During the sea-breeze project based at Lasham, 45 km inland, over a period of 12 years it was suspected that fronts much nearer the coast might retreat more frequently than those which reach Lasham. This was confirmed by work from

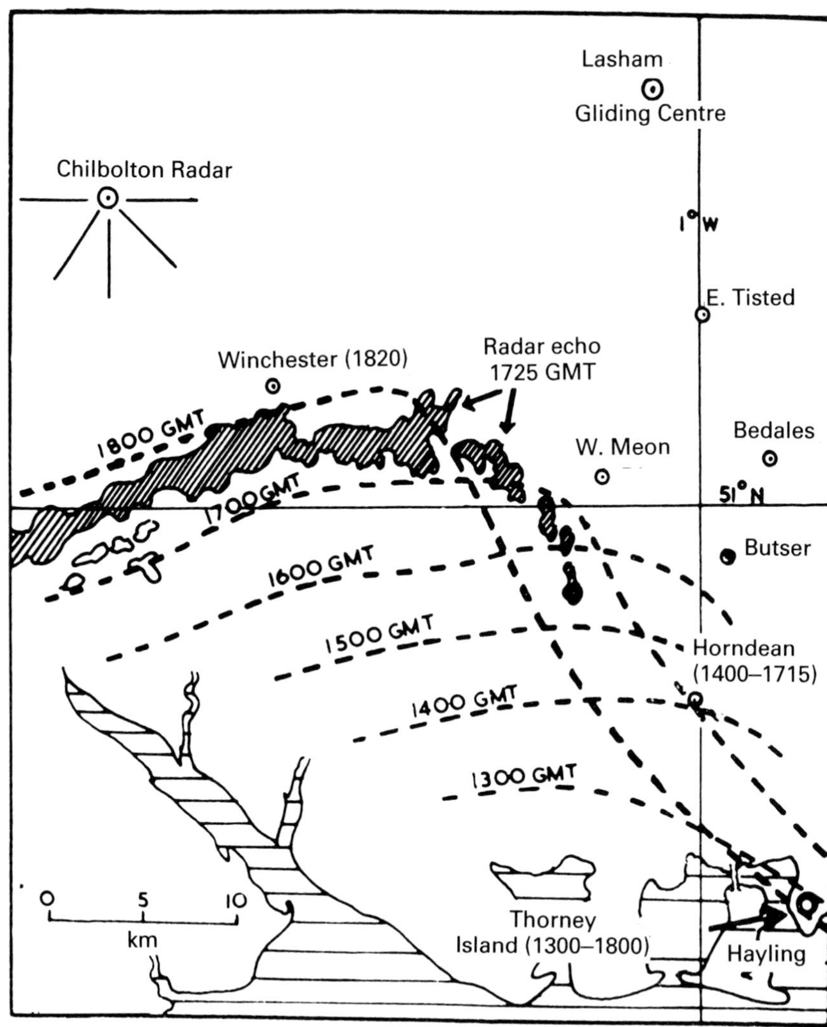

Figure 3.14 Part of southern England on 5 June 1973, showing how an increasing easterly wind was able to distort a sea-breeze front and force it to retreat seaward. Dashed lines show hourly positions of the front, shading indicates radar echo at 1725 GMT.

an extended lake-breeze project, lasting 6 years, carried out from the east coast of Lake Michigan (Ryznar & Touma, 1981).

During the 6-year period, 187 lake breezes were recorded, of which a total of 76, nearly half, moved as far inland as the 19 km station. As many as 50 lake breezes had moved some distance inland but then retreated lakeward, 24 of these returned as far as the shoreline itself. This was usually due to an increase in offshore wind speed, but it was sometimes associated with an increase in cloudiness.

3.6 Sea-breeze undular bore

A change in the form of the sea-breeze front in the evening has been observed in Australia, where it was described by Clarke (1965) as a vortex forming at the front and separating from the main cool flow which followed it. In southern England, glider pilots flying in the upcurrents at the sea-breeze front also noted, as sunset approached, a fundamental change in the nature of soaring conditions. The characteristic narrow line of turbulent rising air, typical of a normal sea-breeze front, changed its nature until at sunset it had become a very smooth area of rising air extending along a band a kilometre or more wide. The extreme smoothness recalled the conditions found when soaring in atmospheric waves. It was also sometimes found that this area appeared to be cut off from the following sea breeze.

The second aeroplane flight shown above in figure 3.11, which was completed at sunset, 1915 GMT, showed a change in the form of the front, which appeared to have a cut-off head. This is now recognised as an early stage of an atmospheric bore formed by the advance of the sea-breeze front through the stable layer of the evening inversion. This inversion is a layer in which the normal temperature gradient with height is inverted, due to rapid cooling from the ground in the evening.

The formation of a tidal bore in a river is a familiar occurrence, a well-known example appears in the River Severn in England, but the formation of an internal bore at the interface between two layers of air of different densities is much less well known. It has been shown that an atmospheric gravity current can cause a bore to form in a stable atmospheric layer (Rottman & Simpson, 1989), and a good example is the movement of the sea-breeze gravity current through an evening inversion layer. Figure 3.15 is a simplified display of conditions meas-

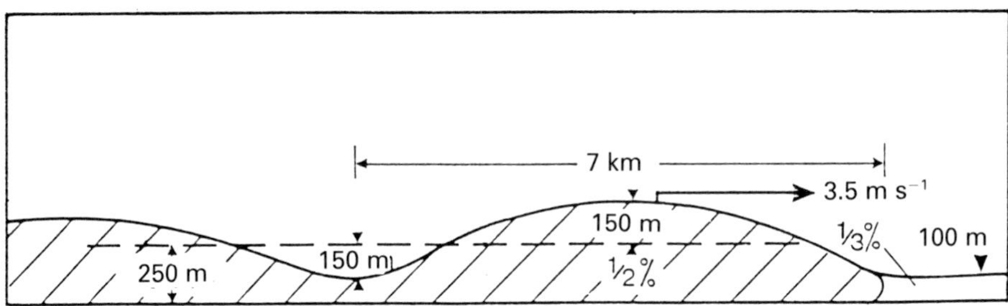

Figure 3.15. An early stage of the formation of a bore-wave by the sea-breeze front moving (from the left) through the evening inversion layer measured north of Lasham, southern England, on 14 June 1973. The shaded area is the sea breeze and the dashed line shows its mean height.

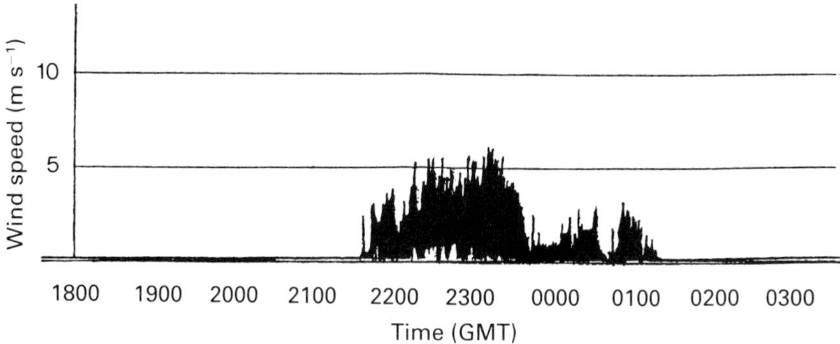

Figure 3.16. Record of wind strength at Harwell, southern England, 85 km from the coast, during the night of 14 June 1973. Variations in strength appear as waves pass in the sea-breeze bore.

Figure 3.17. Roll-cloud marking the arrival of the Morning Glory at Burketown, northern Australia, at 0630 local time on 12 October 1980. (Photo by Roger Smith.)

ured during the second aeroplane flight of 14 June 1973, showing the early stages of the formation of a sea-breeze bore-wave. This inversion had reached a thickness of 100 m, and had a mean density only a little less that of the sea breeze. The wavelength of the undular bore was about 7 km.

Later in the evening more than one wave was apparent and figure 3.16 shows such a record in which the bore is displayed by rhythmic variations of wind strength as each wave passes.

3.61 The Morning Glory

The atmospheric bores described above, which are formed by sea-breeze fronts, sometimes travel hundreds of kilometres. They propagate in stable atmospheric layers which form at night, and may still exist the following morning.

The 'Morning Glory' is the most spectacular of these; it occurs in northern Australia near the southern coast of the Gulf of Carpentaria. A series of roll-clouds, extending right across the sky, arrives from the east soon after sunrise, accompanied by a wind-squall and a sharp rise in surface pressure (Smith & Goodfield, 1981). A typical example is shown in figure 3.17.

The origin of the Morning Glory has been thoroughly investigated and it appears to be formed from the collision of the two sea breezes from the east and west coasts (Clarke, 1984). The way in which bores are formed from collisions is described later in section 11.6.

Atmospheric bores marked by roll-clouds similar to those of the Morning Glory have been observed world-wide; their most common origin seems to be cold outflows from thunderstorms.

4

Sea-breeze forecasting

The cause of the sea breeze is the different rise in temperature between the land and sea surfaces during the day; forecasters must estimate the expected value of these temperature differences. How a sea breeze develops is strongly influenced by the winds in the large-scale weather systems.

The degree of instability of the air above the land is also significant, but the most favourable conditions for the formation of a marked sea breeze and its penetration inland are not straightforward. The presence of mountains and the shape of bays or headlands can have strong local effects and as a result a procedure for successful sea-breeze forecasting usually turns out to be specific to the region where it has been developed.

4.1 Land and sea temperatures through the year

Comparison of the mean monthly sea temperature at a coastal site with the mean maximum and minimum temperatures over the land can give useful sea-breeze probabilities. Figure 4.1 gives such a record for Thorney island on the south coast of England. From this chart we can see that the land–sea temperature difference is greatest from April to September, and sea breezes will be most likely then. The sea will be on average warmer than the land from October to February, when sea breezes will be less likely.

4.11 Temperature difference required for sea breeze

A classic series of tests were made at Thorney Island over a period of five years to see whether the sea breeze did not blow until the land temperature excess

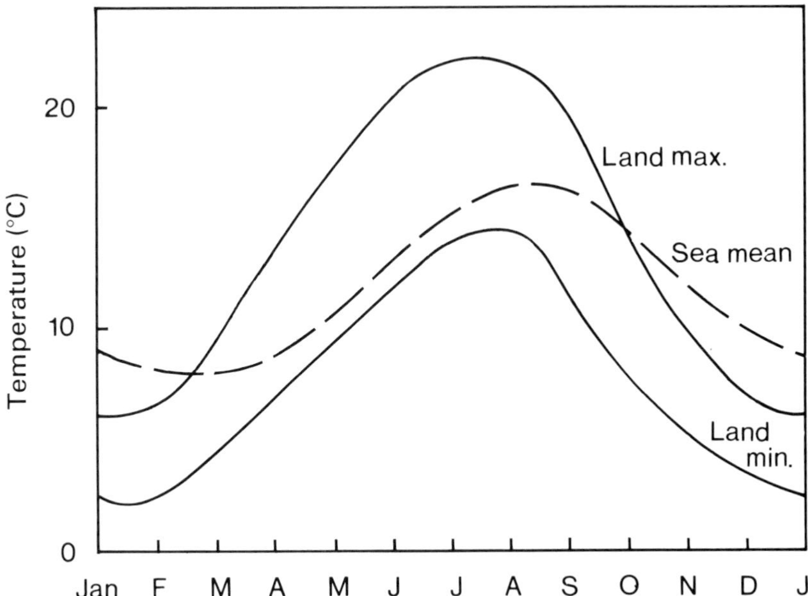

Figure 4.1. Temperatures of land and sea at Thorney island during 1952. Sea breezes are most frequent from March to September, when the land is warmer than the sea.

reached a certain value, and how this value depended on the strength of the opposing wind.

The wind at 1000 m was estimated from ascents at points inland and the observations were divided into three groups according to this wind direction. It was seen at once that a critical line could be drawn dividing occasions when a sea breeze did occur from those when it did not (see figure 4.2) (Watts, 1955). This curve provided a simple, reliable method of forecasting whether or not a sea breeze would occur at this site.

4.2 Sea-breeze index

Dimensional analysis yields a ratio that is representative of the balance of the forces which control the establishment of a sea breeze.

The inertial force, $\varrho U^2/2$, and the buoyancy force, $\varrho g \beta \Delta T$ (where ϱ is density, U is wind speed, β is specific heat and ΔT is temperature difference) when taken as a ratio, give the dimensionless number $U^2/2g\beta\Delta T$, but g is constant over the small vertical extent of the sea breeze, and β is constant over the range of temperatures involved. For this reason the 'sea-breeze index $U^2/\Delta T$' is representative of the ratio of the controlling forces, and its value should give some idea

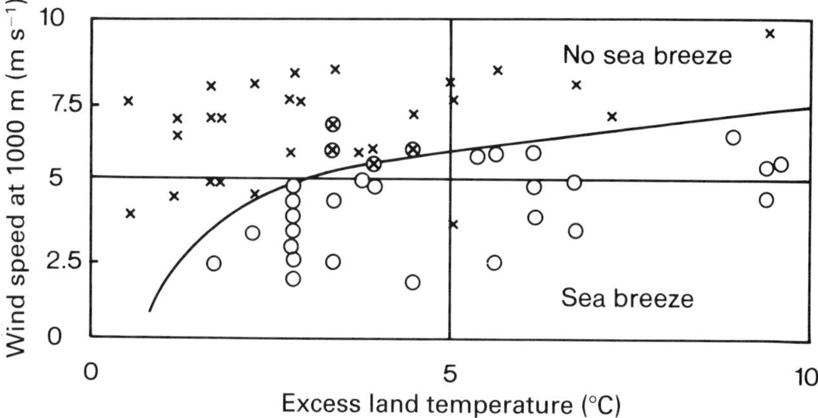

Figure 4.2. The range of wind speed and temperature differences for which sea breezes did or did not occur at Thorney Island. O, sea breeze; x, no sea breeze; ⊗, marginal case. (After Watts, 1955.)

of the likelihood of the appearance of the sea breeze. For large values of the ratio U^2 predominates and no sea breeze will advance against the wind. For small values the temperature difference is large and a sea breeze would be expected.

T and U are obtained by selecting a site uninfluenced by sea-breeze effect and therefore representative of temperature and wind velocities for the local area. ΔT is simply the difference between land and sea-water temperature.

Measurements made at the north-east tip of Lake Erie, where it is about 30 km wide (Biggs & Graves, 1962), gave a critical value for the 'lake-breeze index' of 3.0. If a transition zone between 2.7 and 3.2 was recognised, then the procedure had an accuracy of 97% in the test period.

To obtain a 'sea-breeze index' for Thorney Island a new graph has been derived from the original measurements, plotting the square of the opposing wind against the temperature difference. From figure 4.3 it can be seen that the sea-breeze index has a value of 7. This is larger than the value obtained for Lake Erie, and can be partly accounted for by the difference in the methods of selection of the wind speed U. The wind for the Thorney Island forecast was the value at 1000 m from a pilot balloon ascent, and would be expected to be give greater values than the Lake Erie index, which was based on surface wind speed at a nearby site unaffected by the lake breeze.

Since this method offers high accuracy for comparatively little effort, a series of measurements of the lake-breeze index was made for the lake breeze at Chicago (Lyons, 1972). This time the geostrophic wind over the southern basin of the Lake was employed. The value of ΔN, the 4 mb isobar spacing in km was measured from the 0600 CST surface chart. The lake-breeze index was then

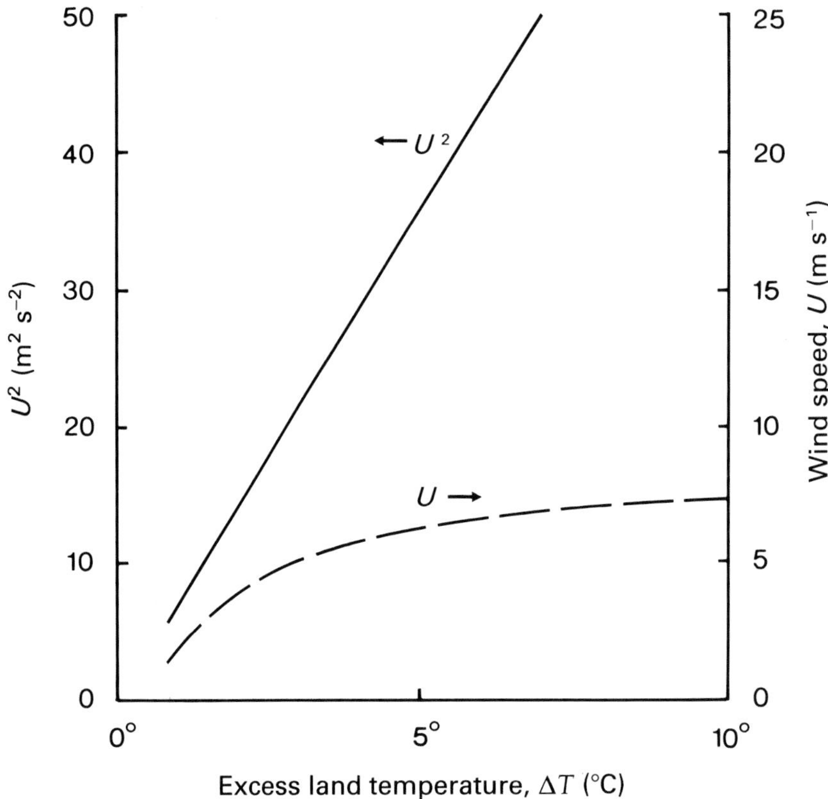

Figure 4.3. The dividing line for sea-breeze occurrence at Thorney Island obtained from figure 4.2 by plotting the square of the wind speed, U^{-2}. The result is a straight line, through the origin, on which the value of $U^2/\Delta T$ is 7 m^2 s^{-2} °C^{-1}.

(12.71×10^6) divided by the product of ΔN^2 and the estimated temperature difference. The critical value of the index here was found to be 10.

4.3 Prediction of inland penetration

In some parts of the world, notably in Australia, the sea breeze has been observed to penetrate inland for distances of 200 km or more. In Britain, with nowhere much more than about 100 km from the coast, extended studies have been made of the sea-breeze onset at several sites about 50 km from the coast and some effects have been recorded as far as 100 km inland.

The main point of agreement of all these studies has been that the overall wind strength of the day is a very important factor and with an offshore wind greater than 5 m s^{-1} deep penetration is unlikely. This applies no matter how large the expected temperature difference, which, however, needs to be at least 5 °C.

The other factors are not so clear, but are all related in some way to the nature and scale of the convection expected during the day. Figure 4.4 shows the critical wind strength and direction required for the sea breeze to arrive at Scampton, 50 km from the east coast (Pepperdine, 1966). Convection is a prerequisite, but must not be too deep, cumulus tops must not be greater than 3000 m.

At Manby, also in Lincolnshire (Brittain, 1966), if convection is less than 1800 m, the sea breeze is not observed, and the offshore wind is again considered more important as a control than temperature difference.

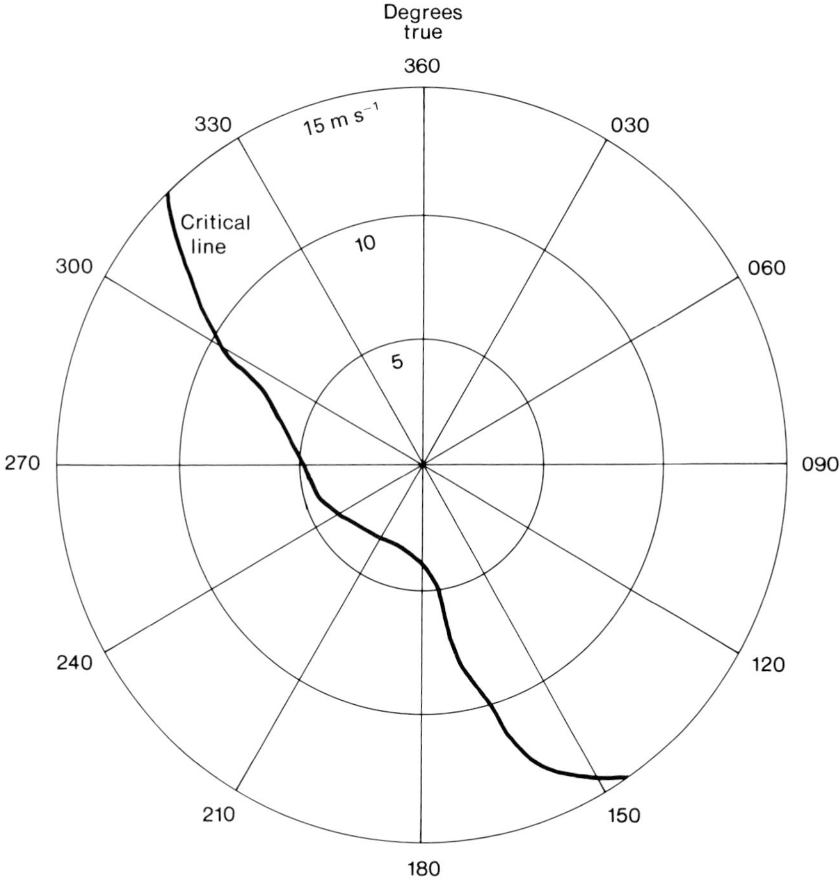

Figure 4.4. Critical line for the occurrence of the sea breeze at Scampton, Lincolnshire. If the surface wind at 0900 h lies to the south-west of the critical line, sea breeze will not affect Scampton. (After Pepperdine, 1966.)

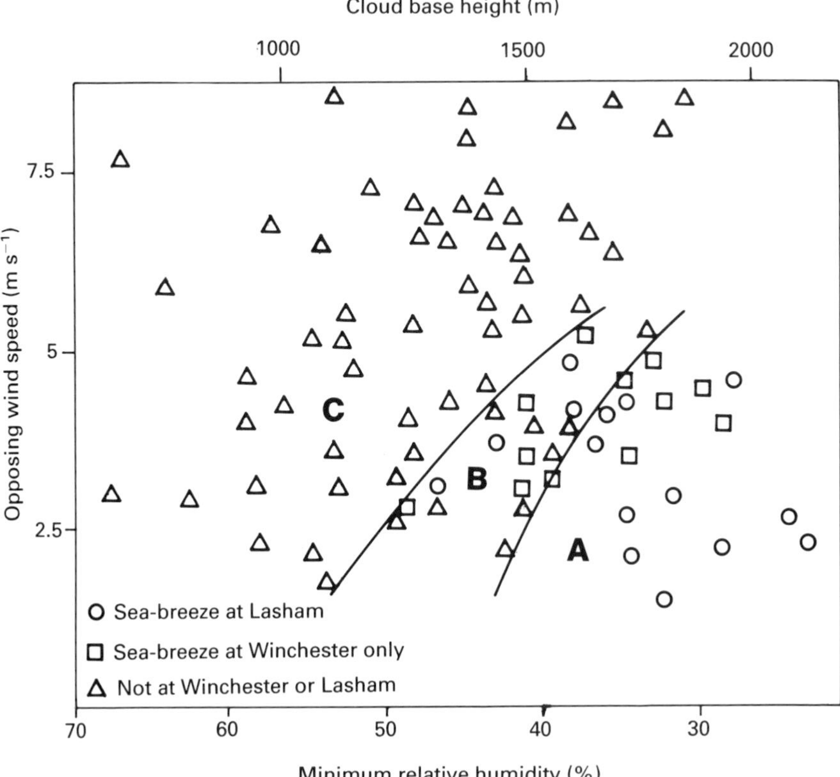

Figure 4.5. The range of opposing wind speed and minimum relative humidity for which the sea breeze did or did not reach Winchester, 30 km inland, and Lasham, 50 km inland. ○ sea breeze at Lasham, □ sea breeze at Winchester, △ sea breeze at neither. Sea-breeze forecast for Lasham would be: (A) likely, (B) possible, (C) very unlikely.

At the Gliding Centre at Lasham, 50 km from the south coast of England, where the sea-breeze arrival seriously modifies soaring conditions, the arrival of sea-breeze fronts was found to depend on the opposing wind and the height of cloudbase (Simpson, 1966). The latter is easy for any glider pilot to see during a soaring flight, and can be simply monitored from the ground by measuring the relative humidity. Figure 4.5 shows a plot of sea-breeze days at Thorney Island (a prerequisite for sea breeze at Lasham) in which the opposing wind is plotted against the lowest value of relative humidity attained, from which the convective cloud base is easily calculated. It appears that 5 m s^{-1} is the wind limit, and that relative humidity needs to be below 40%, i.e. with the cloud base above 1800 m.

4.4 Forecasting from the state of the tide

The temperature of tidal flats shows a periodic behaviour due to flooding which is different from the periodicity caused by solar heating. Also, the inflow and outflow of water will influence the surface temperature of the sea in the tidal area. The influence of the tidal effect on the ground thus depends on the phase of the tidal cycle, that is, on the time of day when the tide is high.

The frequency of inland penetration of the sea breeze in southern England over a 12-year period has been studied as a function of the tide (Simpson, Mansfield & Milford, 1977). A strong dependence of this frequency on the phase of the tidal cycle was found. These results are illustrated in figure 4.6, which shows the frequency of Lasham sea breezes (50 km inland) as a function of time of high tide at Hayling Island, on the coast. A test-fit of the results to this sine curve showed the results to be significant at the 1% level. The unexpected strength of the result was not fully explained, but it was suggested that it was due to the large area of sand and mud-flat near Thorney Island over which the sea breeze had to pass.

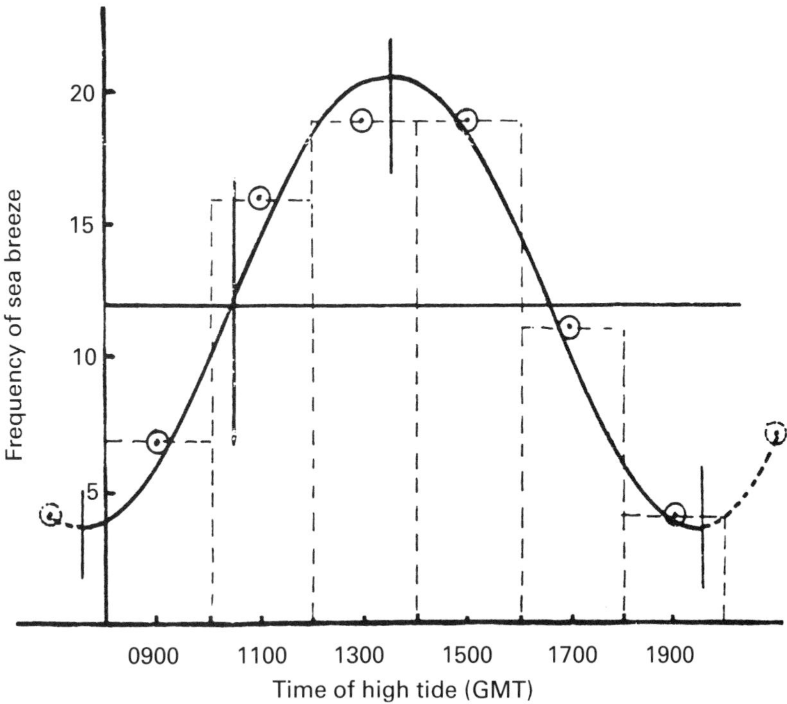

Figure 4.6. Frequency of Lasham sea breeze as a function of time of high tide at Hayling Island. The sea breeze is most likely to occur when high tide is between 1000 and 1600 GMT.

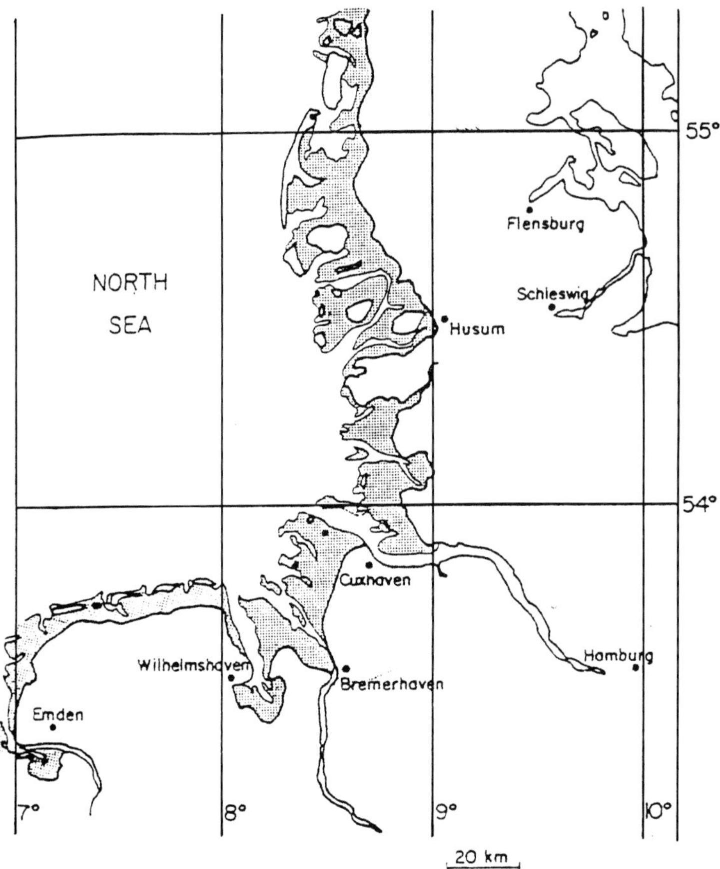

Figure 4.7. Part of the North Sea coast of Germany used in a numerical model of the effect of tidal flats on the behaviour of the sea breeze. The sand-bar areas are shaded. (From Kessler *et al.*, 1985.)

A pioneer numerical study has been performed on the effects of tidal flats (Kessler *et al.* 1985) on the sea breeze. The area chosen was part of the north coast of Germany and figure 4.7 shows the extensive sand bars in the area of study.

Considerable dependence was found of the sea breeze in this area on the phase of the tidal cycle. A flooded tidal flat appeared to act as a sea, whereas an unflooded tidal flat caused the circulation to be much more diffuse.

Comparable studies along the Dutch Coast (Meesters *et al.*, 1989) suggest that this effect is more general. The site chosen was the Island Schiermonnikoop, where large areas of unvegetated sand flats are flooded only at high tide. The

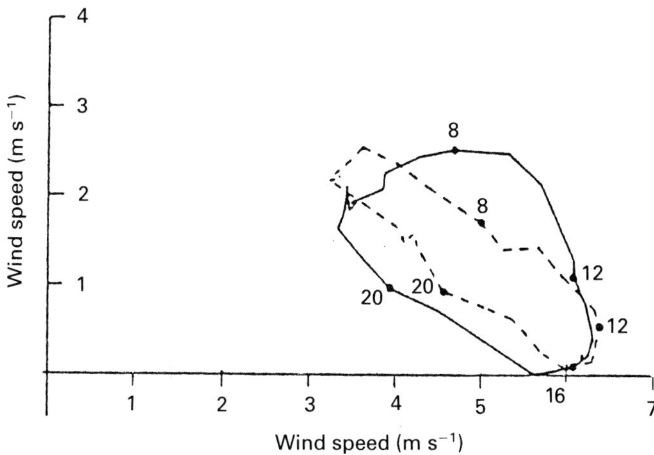

Figure 4.8. Sea-breeze hodographs at the Dutch Coast with sand flats averaged over groups of days with the wind in the west quandrant during July and August. They show two curves:

——— Group I, days with high tide between 0330 and 0930 UT.

– – – – Group II, all other days.

hourly mean values of wind direction and wind speed at a height of 6 m at Groene Glop on the island were measured over a period of ten years.

Mean hodographs were plotted to show diurnal averages over groups of days with wind constantly in one quadrant. For days in group I the tidal flat is unflooded roughly at those hours for which the sun's heating is strongest, for days in group II, it is flooded at precisely these hours. The idea behind this subdivision was that, as far as the heat balance was concerned, at daytime the behaviour of the tidal flat will be land-like for days in group I and sea-like for days in group II.

Figure 4.8 shows hodographs with westerly winds. A dependence of wind speed with tides is found, and the tide dependence is seen to be mainly a consequence of the tidal oscillation with the heat balance on the tidal flat, but the full explanation remains very complex.

In conclusion, there is little doubt that at coasts with large tidal mud flats the appropriate use of the state of the tide can be of use in sea-breeze forecasting.

5

Other local winds

Two main classes of local winds can be distinguished from each other. The first are thermally-induced winds. These are generated by differential heating of parts of the Earth's surface, of which the sea breeze is only one example. They are found especially in mountainous country.

The second class consist of orographic winds which occur when certain vigorous large-scale weather systems combine with local topography to produce strong winds. These have distinct local characteristics; examples are the cold 'bora' and the hot 'föhn', with counterparts in many countries.

5.1 Winds from diurnal heating on mountains

Many complex systems of thermally induced winds on mountains and valleys have been studied (Atkinson, 1981; Barry, 1961). In order of increasing size we can recognise: (*a*) the wind formed on individual slopes on a mountain or valley; (*b*) the effect of the slopes generating the larger 'along-valley' winds, often called 'mountain and valley' winds; and (*c*) a larger system, which has been called 'mountain–plain' winds, with influence extending on a scale of 100 km or more.

5.11 Up- and down-slope winds

Winds formed on mountain slopes form part of a thermally caused diurnal circulation, similar to that of the land-and-sea breeze. Figure 5.1 shows the distortion of the horizontal lines of potential temperature caused in (*a*) by surface heating during the day and (*b*) by radiative cooling at night. The horizontal temperature differences result in a rising air current in the day, called an anabatic flow and a descending one at night, called an katabatic flow. Measurements with pilot-balloons have shown the depth of these flows to be about 30 m and the strength to be up to about 3 m s^{-1}. Figure 5.2 shows results of some observations made on the slopes of the Nordkette, Innsbruck (Defant, 1949). Very similar results have also been found in observations in south Australia.

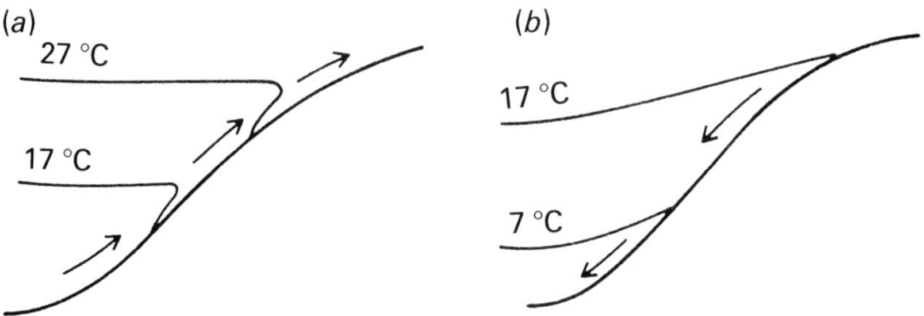

Figure 5.1. Distortion of lines of potential temperature on slopes during (*a*) heating and (*b*) radiative cooling. (After Cramer & Lynott, 1961.)

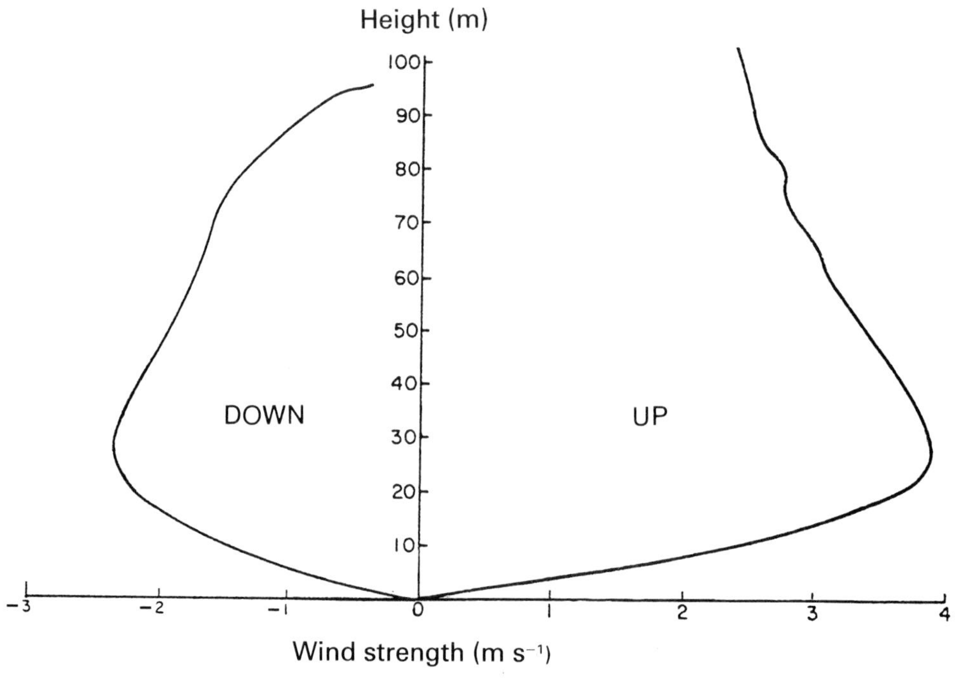

Figure 5.2. Strength of slope winds at different heights above the ground. From pilot-ballon observations on the Nordkette, Innsbruck. (From Defant, 1949.)

Although the upslope winds are not usually very strong, their presence can be deduced from the formation of cumulus clouds on sunny mountain slopes (see figure 5.3).

Figure 5.3. Growth of cumulus clouds on heated mountain slopes

These anabatic currents up slopes can be used by glider pilots in mountain soaring. However, pilots need to fly very accurately to obtain this 'lift', since the maximum strength may occur only two wing-spans from the mountain face. Figure 5.4 shows three gliders soaring close to a steep rock face which is being heated by the sun.

The katabatic, or down-slope winds, have attracted much more attention. Observations have supported a hydraulic approach to include dependence on time of flow and the main control being turbulent mixing across the interface with the ambient air. (Manins & Sawford, 1979). This view is also supported by experimental laboratory experiments on gravity currents down slopes.

On ice-covered slopes, the diurnal regime of anabatic winds may be weak or even absent, with strong katabatic winds blowing throughout day and night. Very high speeds occur on the katabatic flows on the margins of the Greenland and Antarctic ice sheets. Figure 5.5 shows a recording of the onset of Antarctic katabatic wind at Mawson (Streten, 1965), where it can be seen that three hours after the onset a gusty wind of between 15 and 20 m s^{-1} is still blowing. These descending winds can reach such very high speeds partly because the air moves in a relatively shallow layer in 'shooting' instead of 'tranquil' flow. The former type of flow is called 'supercritical' since it is faster than the maximum speed

Figure 5.4. Gliders soaring on the sun-lit rock face at Gache in France. (Photo by Dr. Claus Dieter Zink.)

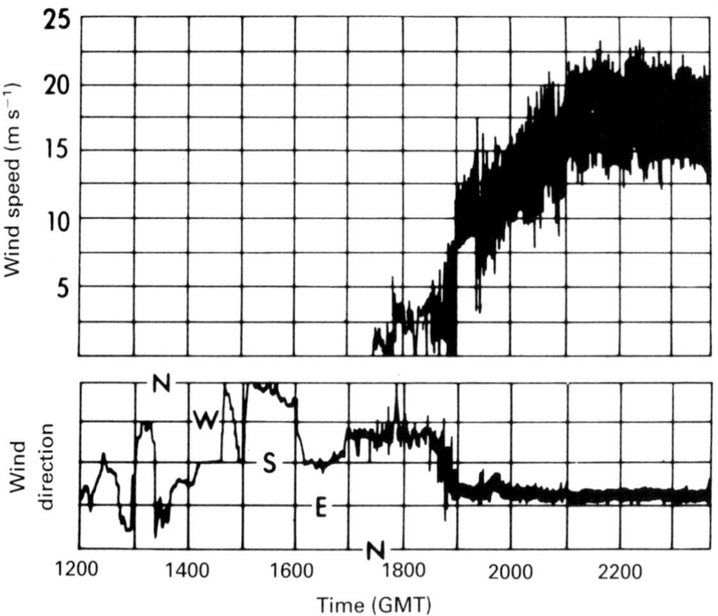

Figure 5.5. Wind speed and direction at the onset of a typical katabatic wind, 21 August 1960. Mawson, Antarctica. (After Streten, 1963.)

which waves can travel on the surface. This class corresponds to the much discussed 'supersonic' flow of air moving at speeds faster than normal sound waves.

When the air returns from supercritical shooting to subcritical tranquil flow, the height of the colder surface air increases rather suddenly with a marked weakening of the wind speed. This 'hydraulic jump' generally occurs not far from the coast and may move forwards or backwards.

The unusual experience of actually walking through a standing katabatic jump has been described by an observer who walked through one, making measurements, near Davis, Antarctica (Lied, 1964). Starting from downhill of the jump he had the odd sensation of approaching a strongly roaring wall of drifting snow, which was neither retreating or advancing, towering 100 metres above him. After taking measurements immediately outside the edge of the jump the observer stepped into a totally different world, like walking through a door opening out into a full blizzard. At the edge of the jump he experienced severe buffeting with violently rotating whirls of wind and snow. At this point measurements showed a sudden drop in pressure and an immediate rise in temperature; to check his readings he repeated his measurements on each side of the jump. On walking back down the slope, he found that when he left the jump the transition from violent rough flow to light or variable conditions usually occurred over a distance of only about 5 metres.

Sets of measurements of pressure, temperature and wind were made through standing jumps on four different occasions. The results on three days were all very close to each other: pressure, 2 mb fall; temperature, 2 °C rise; wind, 10 m s^{-1} increase. However, on the fourth day, the one illustrated in figure 5.6, the changes were all much greater: pressure, 20 mb fall; temperature, 5.5 °C rise; wind, 14 m s^{-1} increase. Figure 5.6 shows the different regions through which it was possible to walk.

NATURE OF ONSET OF KATABATIC WINDS

Many katabatic flows have a peculiar structure in which the wind force rapidly increases from zero to its maximum, and then gradually fades away. The frontogenesis of katabatic winds should have some features in common with the frontogenesis observed in the sea breeze (described in Chapter 3) since the more dense cold air in a gravity current overtakes the less dense air just above it. The occurrence of these sharp fronts has been shown to be more common when an opposing wind is present (Mahrt & Larsen, 1982).

Sleeping in a tent on a mountain is clearly a good way to be made conscious of local katabatic winds, and a report from a mountain slope on Karisimbi north east of Lake Kiwo in Central Africa describes how the observer's tent was almost swept away by an 'air avalanche' (Kuttner, 1949).

Another feature frequently observed in katabatic winds is a series of regular

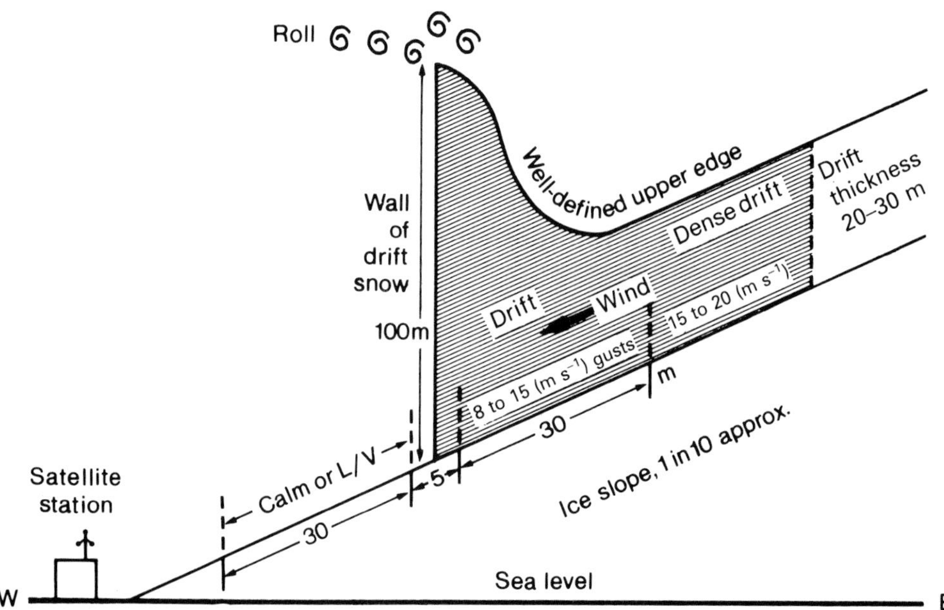

Figure 5.6. Details of a standing jump in a katabatic wind as seen by an observer who was able to walk through the roaring wall of drifting snow from the calm zone lower down the slope. (After Lied, 1964.)

oscillations following the initial front, which has been aptly named an 'air avalanche'. An example of this rhythmic pulsation was carefully observed by a meteorologist when he bivouacked for two nights during a climbing tour on the northern wall of the Hollentaal peak in the Wetterstein Mountains (Scaetta, 1935). The onset at 2111 h was followed by gusts at regular intervals of 5 minutes. This disturbed his sleep regularly throughout the night until 6 o'clock in the morning. The same 5-minute period was kept up the following night.

These variations in wind speed may be associated with periodic separation of the flow from the surface boundary. When this occurs a reverse eddy is found which grows and is eventually shed downstream, enabling the main flow to return to the surface.

5.12 Mountain and valley winds

The diurnal variation of surface heating and cooling generates circulations up and down the sloping floors of valleys as well as up and down the sidewalls.

Figure 5.7 is a simple diagram illustrating these winds. The slope winds form first, for example, the anabatic winds up the slopes are followed by winds up the valleys, called the 'valley winds'. With the onset of radiative cooling in the

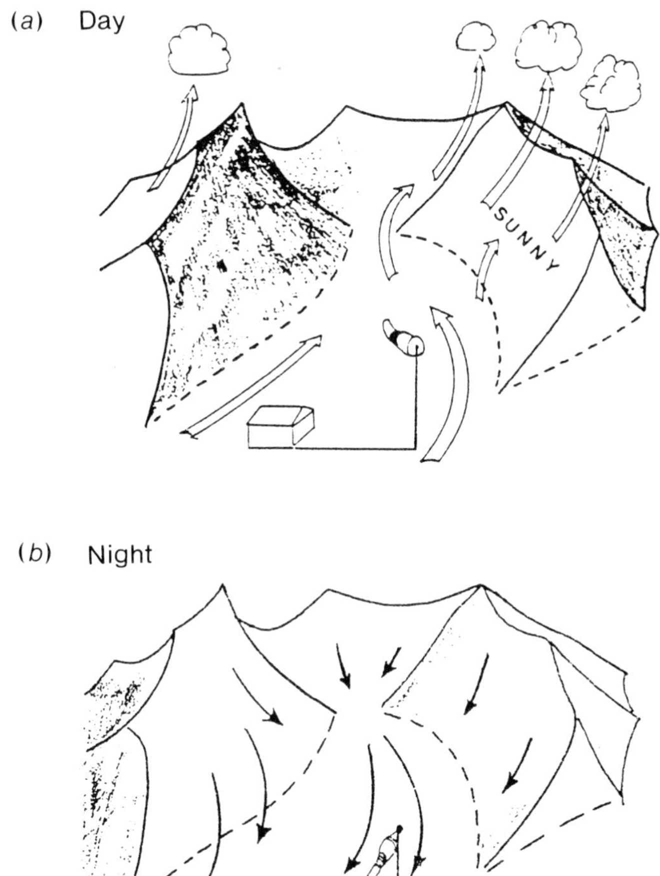

Figure 5.7. The development of valley and mountain winds. (After Bradbury, 1989.)

evening, cold air begins to drain down the slopes eventually leading to a flow down the valley floor towards the valley outlet. This is called the 'mountain wind' and may be maintained all night by the drainage of cold air down the sides of the valley.

Observations suggest that the valley wind usually sets in at the same time along the length of a valley, but the onset of the cold drainage mountain wind occurs at a local front of cold air moving down the valley, as shown in figure 5.8. It is believed that at some places (Thompson, 1984), well-defined surges of cold drainage air only occur when strong opposing winds exist, but strong head winds do not seem to be essential everywhere for the formation of these fronts.

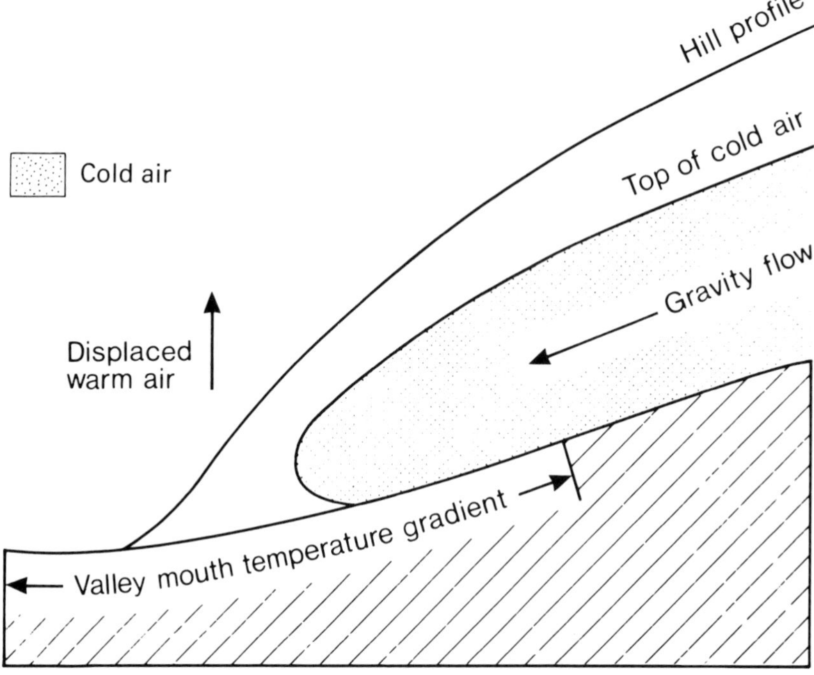

Figure 5.8. Initial stage of katabatic development as a minor cold front moving down a slope. From measurements at the mouth of the Red Butte Canyon. (After Thompson, 1984.)

Oscillations, as already described in smaller-scale down-slope winds, are often very marked in mountain winds. Figure 5.9 shows some wind velocity profiles measured in the mountain wind near Pietermaritzburg in South Africa, (Tyson, 1968) which show the wind strength more than doubling in a period of 20 minutes during such a surge. Other measurements in India (Atmanathan, 1931) showed roughly the same periods.

It has been suggested that these surges are due to adiabatic warming of the air during its descent, causing the wind to decelerate. When radiative cooling has built up a sufficient temperature gradient, further acceleration would occur and this could repeat itself several times during the night.

In complex mountain terrain the mountain winds formed from separate peaks may combine as they descend towards the plain. Figure 5.10 illustrates the results of some field work in which the gravity currents descending from Alpine mountains combine as they flow down towards Lake Maggiore in North Italy.

The principal characteristics of drainage flows are preserved as a gravity current while the flow moves away from the foothills over the plain. For example, such progress has been observed over 25 km of gently undulating ground

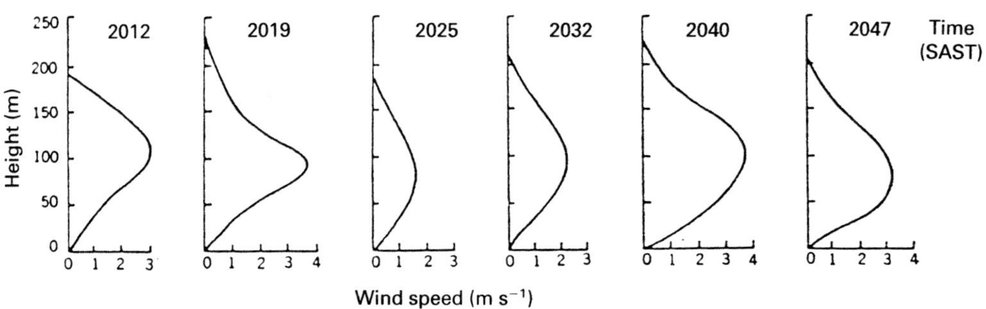

Figure 5.9. The occurrence of a surge with an interval of 20 minutes in the mountain wind at Pietermaritzburg in South Africa on 5 July 1965. (Thompson, 1984.)

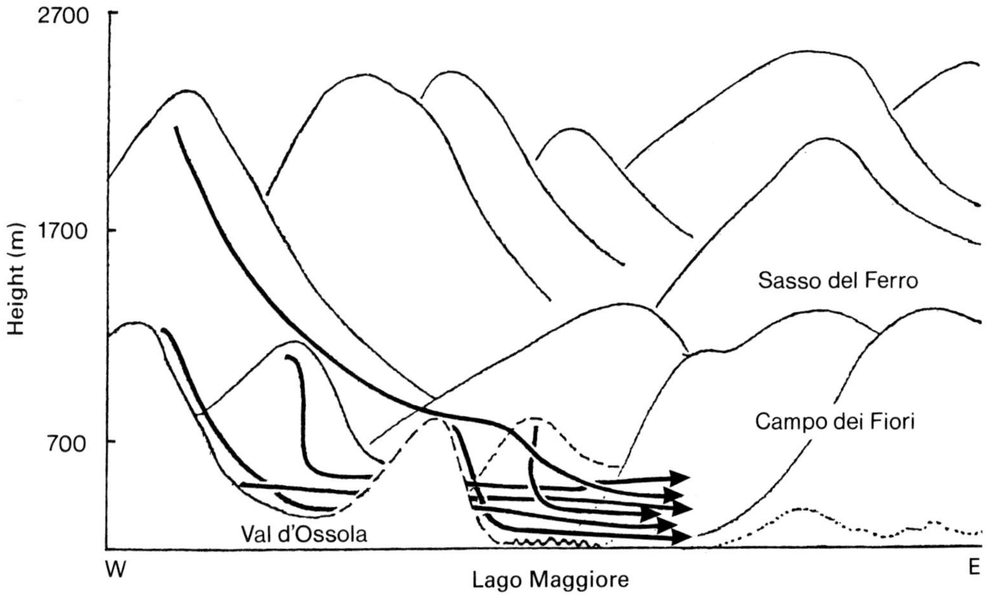

Figure 5.10. Section from west to east showing gravity currents descending from Alpine mountains to the level of Lake Maggiore.

between the foothills of the Rocky Mountains to Boulder, Colorado where precise measurements have been made on many occasions (Blumen, 1984). An acoustic sounder (sonar) trace in figure 5.11 shows a typical example with a sharp frontal onset at about 1920. Three separate surges can be detected in this trace, with a period of about 45 minutes. Observations made here are in general agreement with gravity current measurements made in laboratory channels (to be described in Chapter 10).

Figure 5.11. Acoustic sounder trace showing the arrival of drainage flow at Boulder, Colorado on 8 October 1980. (Courtesy of W. Blumen.)

5.2 Orographic winds

Vigorous weather systems may combine with local mountain topography to produce strong down-slope winds with very distinctive local characteristics. Two well-known examples are the *bora*, a down-slope cold wind and the *föhn*, or hot down-slope wind. Similar winds occur locally in many parts of the world, having an appreciable influence on human life, health and mood.

The bora is one of the strongest winds, sometimes of hurricane force; it is strong enough to overturn trains and blow roofs off houses. When this cold wind is accompanied by freezing rain it destroys power cables and poles. One of the most famous and well-documented is the Yugoslavian bora in the mountains along the Adriatic coast.

The föhn was first studied by meteorologists in the Swiss Alps, where it has the reputation of causing feelings of depression and increasing the suicide rate on days when it blows. This hot wind may cause sudden melting of thick snow, with consequent risk of avalanches and disastrous floods.

The simple model shown in figure 5.12 will help to make clear the difference in principle between these two different kinds of down-slope winds, one cold and the other hot. The figure shows a simplified atmosphere consisting of two layers, a warm layer above a cold layer of depth h_0, flowing over a mountain of height d_0. Apart from the subcritical flow shown lower left, two very different

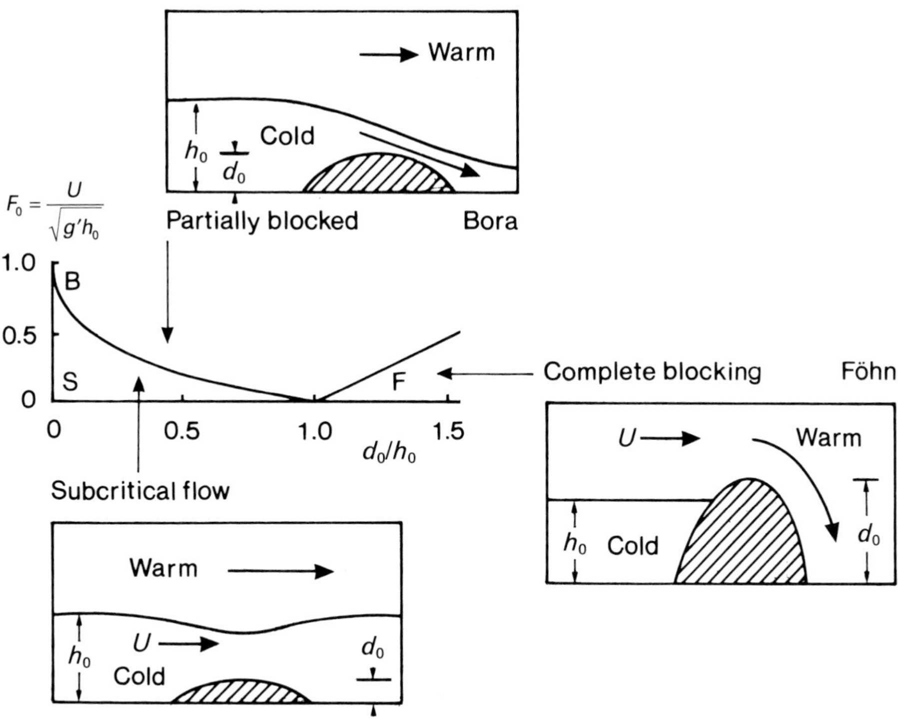

Figure 5.12. Critical conditions in the hydraulic theory of two-layer flow over a mountain. In the bora the cold air is only partly blocked and spills over the mountain in a thin, rapid descent. In the föhn the cold air is completely blocked and the layer of warm air descends, becoming even hotter.

types of flow regime exist. The theoretical critical curves on the left show how the different types of flow depend on the flow speed U (given as a dimensionless Froude number, see section 11.5) and the ratio of the height of the mountain to that of the cold layer.

In the case of the bora the cold air is partly blocked, but spills over the mountain in a thin, and hence rapid, descent. In the föhn the cold air is completely blocked to windward of the mountain and the layer of warm air flows right down to the ground. It becomes much hotter as it is compressed during this deep descent.

Similar models also work well with less simple vertical gradients of temperature in the atmosphere. Also in practice there may be clouds forming on the mountain, which give additional latent heat to the descending air.

5.21 Bora

The main bora region is the Adriatic Coast, from Trieste, Italy to Dubrovnik, former Yugoslavia. Studies made as far back as the nineteenth century, claimed that bora are especially severe where the mountains are higher than 800 m and located within 2 km of the coast.

Other places where the bora prevails in winter in association with the outbreak of cold air from the Eurasian continent are Novorossisk on the Black Sea Coast, and the west coast of Lake Baikal. The Mistral in the south of France is also a bora-type wind and another example is the Helm wind in England, formed west of the Pennines. A similar wind called Oroshi occurs on the Japan Sea side of Honshu, Japan. (This name for Japanese fallwinds may include both bora and föhn.)

In the United States an example which has been studied is the Boulder windstorm, on the eastern slopes of the Rocky mountains.

5.22 Relation between bora and vegetation

The deterioration of vegetation in bora country is most serious on slopes up to 500 m above sea level, where the country is a land almost without vegetation. The effects of the wind followed the destruction of the original vegetation by human activity, the strong winds hindered the restoration of vegetation, and overgrazing of goats and sheep and the felling of trees for firewood intensified the effect of the winds.

During some extensive field work on the problems of bora in the Ajdovscina Basin near the Adriatic coast (Yoshino, 1976) it was found possible to chart the actual areas of greatest wind strength by the use of wind-shaped trees as indicators. These wind-shaped broad-leaved trees grow on their lee side but are stunted on the upwind side, as depicted in figure 5.13. They were graded from

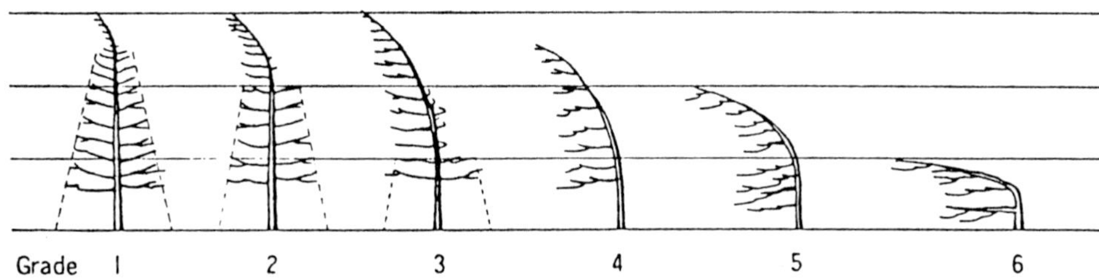

Figure 5.13. Grades of wind-shaped trees in the Ajdovscina Basin, used to estimate the areas of strongest bora. (From Yoshino, 1976.)

1 to 6; grade 0 stands for a symmetrical tree. The wind-distribution map prepared from deformed trees as indicators supported the fact, previously measured from more conventional anemometers, that the wind maxima ran from north-west to south-east. A zone of gentle wind with grade 0 trees appeared just leeward of this axis and further leeward of this weak wind zone a strong wind area existed with grade 3 trees, although this area was small.

5.23 Distribution of houses with stone-laid roofs

Following the observation of wind-shaped trees, the same investigators looked at the distribution of houses with stone-laid roofs in the same area. Stone-laden roofs were classified according to the relative number of stones laid on the roofs, as shown in figure 5.14. The maps prepared from the distribution of these classes of roofs showed areas nearly coinciding with those using wind-shaped trees as indicators, showing the same axis of wind direction and the same strong wind areas. It became clear that stone-laid roofs can be used as indicators of wind conditions, although densely distributed wind-shaped trees are more reliable. These remarkable results make very clear the extremely localised nature of the bora in this district, and its recurrence in the same places.

5.24 Airborne measurements of bora

The first aircraft observations of the bora in Yugoslavia were carried out during the ALPEX project in 1982 (Smith, 1987). The results showed upstream descent and acceleration beginning where the mountains rise. A region of intense turbulence existed downstream of the descending bora with the formation of a thick mixed layer. Some of these features are illustrated in figure 5.15, which uses measurements taken during a flight on 22 March 1982 when the bora had already been blowing for 40 hours. The potential temperature diagram suggests that most of the stability in the incoming flow was concentrated in an inversion between 3 and 4 km. These contours are not exactly streamlines because of the mixing in the turbulent area, which had the effect of weakening the inversion downstream.

5.25 Föhn

The extensive early work in the Alps has led to the name 'föhn' being applied to any warm, downslope wind. These are known by a variety of local names in mountainous districts all over the world. The 'chinook' east of the Rocky moun-

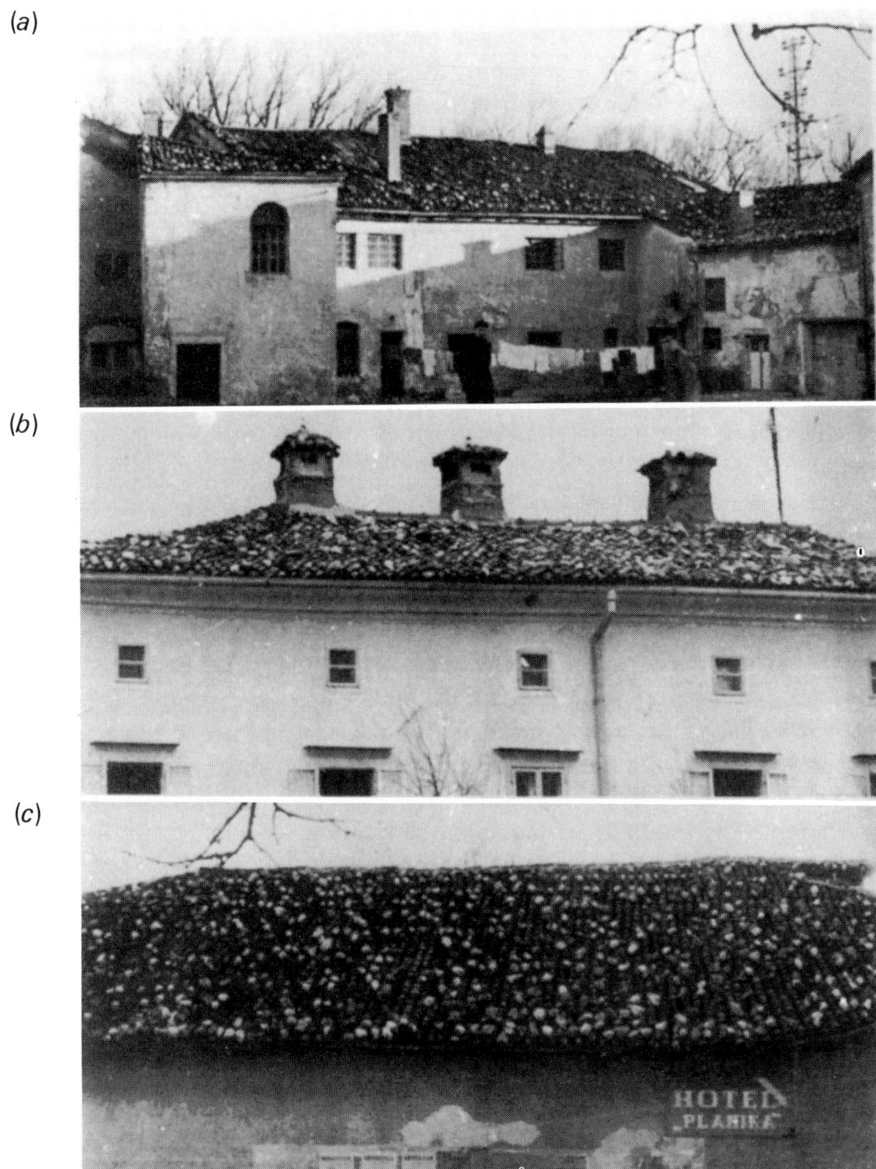

Figure 5.14. Stone-laden roofs of houses in bora country, classified according to the number of stones. The distribution of the classes shown in (*a*), (*b*) and (*c*) was used to estimate the areas of strongest bora. (From Yoshino, 1976.)

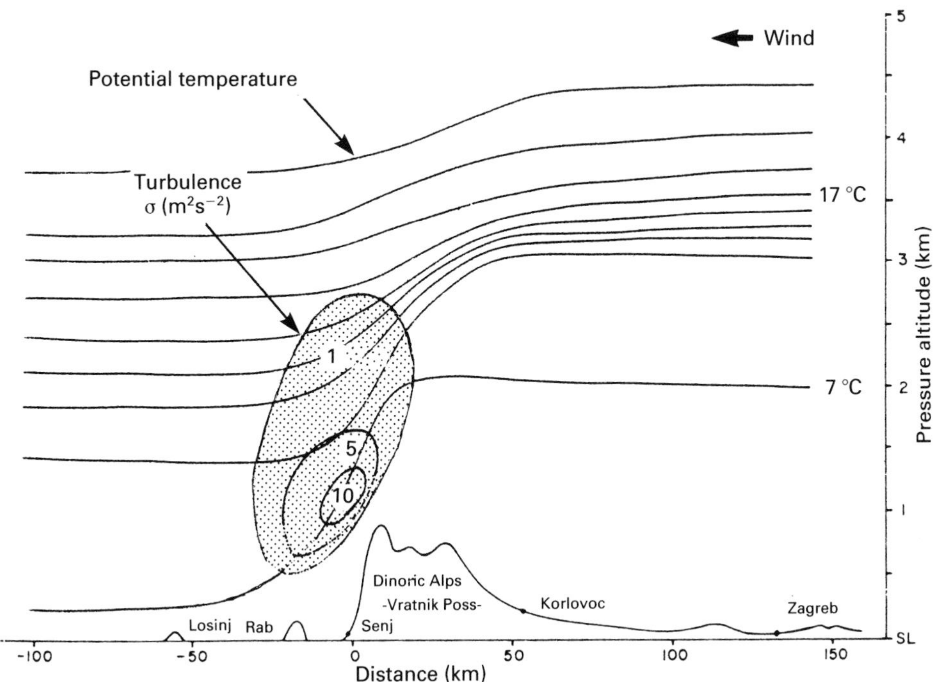

Figure 5.15. Potential temperature cross-section measured by an aircraft during a flight on 22 March 1982, 40 hours after the onset of bora. The results show upstream descent and acceleration beginning where the mountains rise. The turbulence cross-section is also shown. (After Smith, 1987.)

tains has some of the most violent downslope winds that have been recorded, at Boulder, Colorado. Although they have been called chinooks, some of the really strong winds at Boulder are associated with a drop in temperature which is characteristic of bora rather than föhn.

There exist several different approaches to the definition of 'föhn' since different situations may give rise to föhn conditions.

Föhn may occur in cases where blocking of cold air exists at low levels, as shown in figure 5.16(*a*). In these cases it is sufficient for the warm air above to descend from the summit and undergo dry adiabatic compression, at 9.8 °C km^{-1}.

Another mechanism for producing dry hot descending air is shown in figure 5.16(*b*) in this case moisture is removed by cloud formation on the windward slope. Ascent of moist air on the ridge causes the build-up of cloud and precipitation on the windward slope. In this process latent heat is released by condensation above the cloud base and the rising air cools at the lower lapse rate of between 5 and 6 °C km^{-1}. This dry air will be heated at 9.8 °C km^{-1} in the descent which follows.

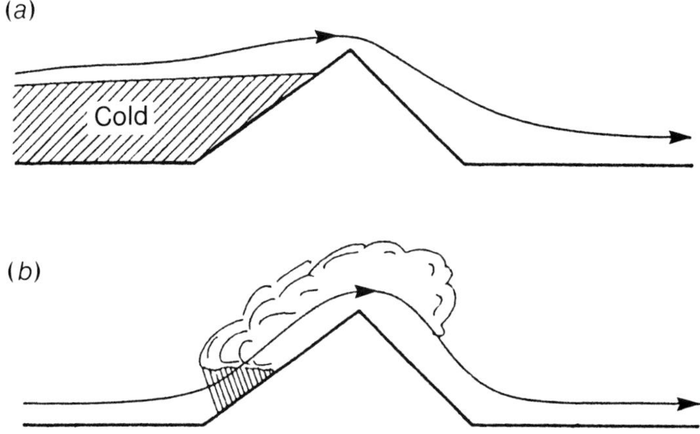

Figure 5.16. Two types of föhn. (*a*) Föhn with damming up of cold air. (*b*) Föhn in a stable atmosphere with strong winds.

5.3 Other local winds induced by differential heating

Other diurnal winds have been detected, closely related to the land-and-sea-breeze, which also result from differential heating (Segal & Arritt, 1992). We have already mentioned the 'lake breeze' as one example of this: large lakes and small seas are difficult to distinguish from each other. Local winds have even been shown to be caused by large rivers.

Other places where differential heating may produce local winds during the day are at boundaries between:

1. bare soil and snow cover: 'snow breeze';
2. sheets of solid cloud cover and adjoining sunny areas: 'cloud breeze';
3. large built-up areas and open countryside.

Local winds are produced by forest fires and have even been observed during total eclipses of the sun.

5.31 The snow breeze

The name 'snow breeze' has been given to the daytime thermally induced flow between the snow and snow-free areas (Johnson *et al.*, 1984).

The presence of snow breeze has not been clearly established through the routine meteorological network, as the irregular snow cover, even over uniform terrain, make its detection almost impossible.

Satellite imagery could be used for measuring surface temperature contrast, and some aeroplane measurements (Segal *et al.*, 1991) have given useful information (see figure 5.17). In most of these cases a synoptic wind of over 10 m s^{-1} was blowing from the bare soil area towards the snow, thus opposing the expected snow breeze, but decreases of potential temperature were measured from the coolest air at 8 °C at a height of 95 m. Decreases in potential temperature of 4.5 °C at 180 m and 1.5 °C at 340 m were recorded, figures comparable to those found in sea-breeze situations.

Well-defined boundaries between snow-free ground areas, under clear sky and light synoptic wind conditions, are common at various locations during the snow season. From the thermal contrasts measured, noticeable snow breezes are likely to occur in those regions.

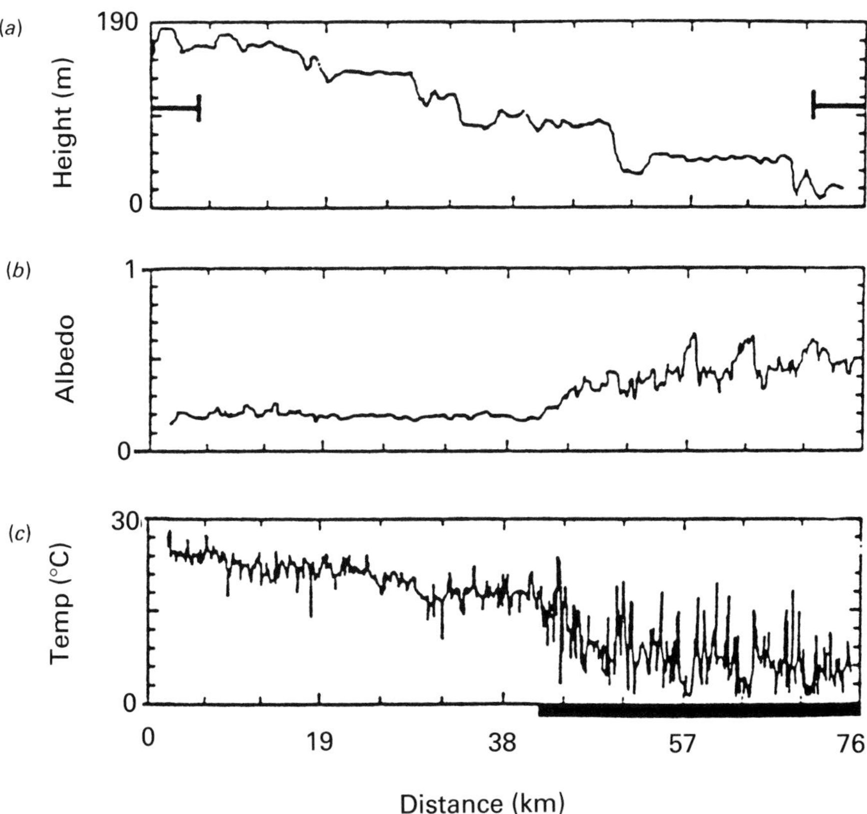

Figure 5.17. Conditions at the boundary of snow-covered ground, which is shown by the dark line. Observational changes in (*a*) ground elevation, (*b*) surface albedo, and (*c*) surface temperature. (From Segal *et al.*, 1991.)

5.32 Cloud breeze, pseudo sea breeze from a 'sea of cloud'

Seen from the air, a layer of stratocumulus looks like a sea of cloud and if this 'sea' happens to have a sharp edge over a land mass it may promote the development of a 'pseudo-sea-breeze' system with a clearly formed front.

This was the situation on 29 April 1958, shown in figure 5.18, when a moist west wind brought inland low stratus over the Irish Sea until it was blocked to the east and to the south-west by high ground (Wallington, 1961). The low stratus was thick enough to prevent a rise in temperature beneath it but to the south, beyond its sharp boundary, bright sunshine warmed the ground and small cumulus were forming. Under these conditions a pseudo-sea-breeze front developed and moved southwards from the edge of the persistent cloud cover: its positions were established from three meteorological stations using records of wind direction, visibility and temperature.

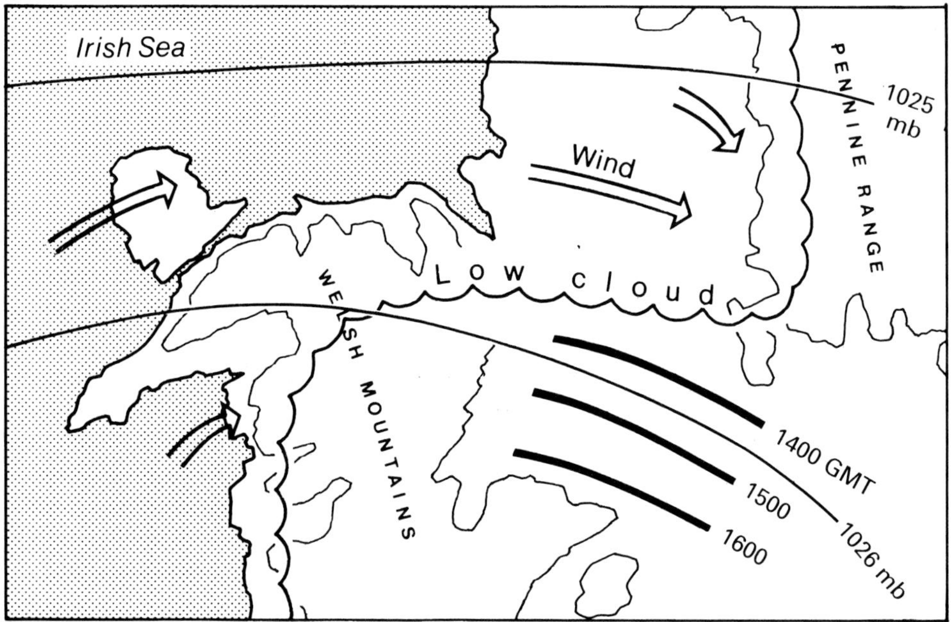

Figure 5.18. Position of a pseudo-sea-breeze front on 29 April 1958 from 1400 to 1600 GMT. The east and south edge of persistent low cloud is marked. (After Wallington, 1961.)

5.33 Cloud breeze investigated by an instrumented glider

Another example has been investigated by an instrumented glider whose crew had originally set out to make measurements in some thermals in south-east England (Milford & Simpson, 1972).

Frontal troughs from the Atlantic had become stationary on 10th September 1971 on a line from Western Ireland to Brest in the English Channel, forming a band of cloud across south-west England. Essa 8 satellite photographs showed that the distribution of cloud on the 10th and 11th September was static, as shown in figure 5.19.

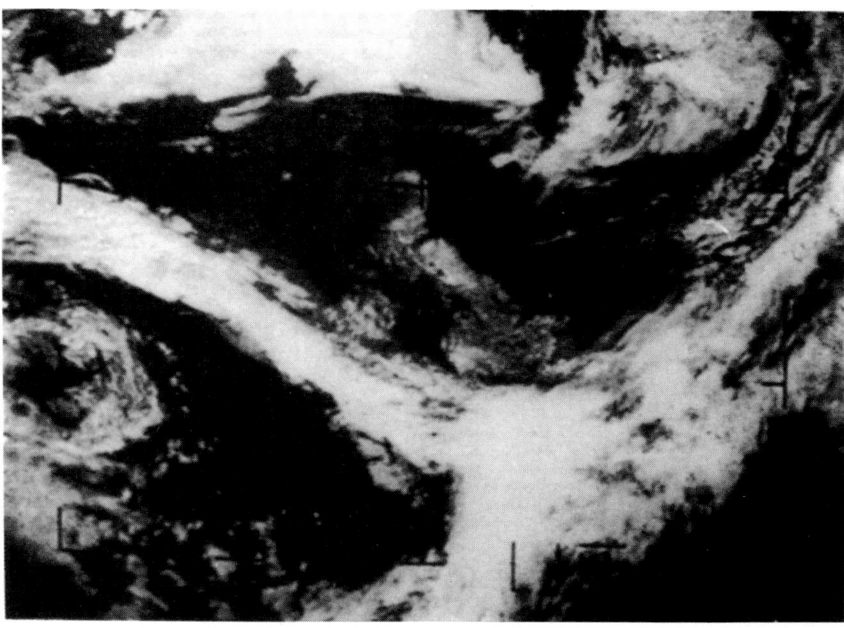

Figure 5.19. Satellite photograph on 11 September 1971, showing area of cloud across south-west England, which had remained there for two days.

On the morning of the 11th the weather was fine to the east of the cloudy zone and surface temperature had risen as much as 12 °C in some places, while in the south-west continuous cloud and heavy rain had kept the land temperature about the same as that of the surrounding sea.

A frontal convergence zone appeared to build up *in situ*: a close parallel to a sea-breeze front which has formed near the coast, but has failed to move inland against an opposing wind. The front was marked by haze and its slope was measured to be 30°. There was a line of ascending air rising at more than 1 m s^{-1}

extending along a 10 km run. Figure 5.20 shows the surface pressure and wind field at 1500 GMT, with the sharp edge of the stationary area of cloud in the south-west. There were also signs of a normal sea-breeze front moving inland to the east, forming an extension of the pseudo-sea-breeze front.

This type of local shearline is probably not uncommon, but only a few of them have been documented. Such convergence zones may be expected to form near the edge of persistent cloud cover with strong surface heating in an adjacent area, having general properties similar to those of sea-breeze fronts. They may

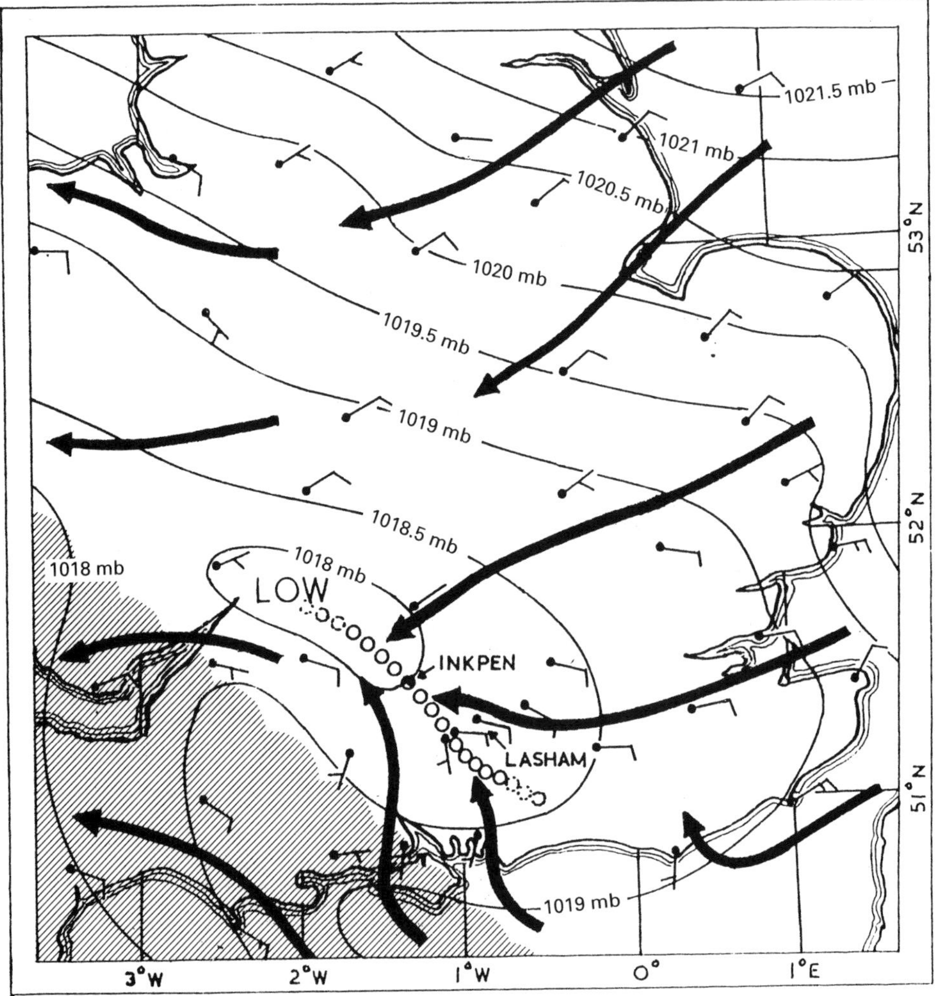

Figure 5.20. Cloud-breeze front, marked by oooooo, observed by an instrumented glider. Surface pressure and winds are shown at 1500 GMT on 11 September 1971. The cloudy area to the south-west is shaded. The heavy black lines show the streamlines which converge towards the marked front.

be significant advectors of pollution and be important in localising shower formation.

5.34 Land and river breezes

There are only a few investigations of land and river breezes, although these effects must be appreciable round large rivers and worthy of more study.

The river breeze is well developed on rivers such as the Orinoco and its effects can be seen in satellite pictures. The clouds above a uniform area of tropical jungle in the Orinoco basin are shown in figure 5.21. Here, about 30% of the sky is filled with separate cumulus clouds, two or three kilometres apart.

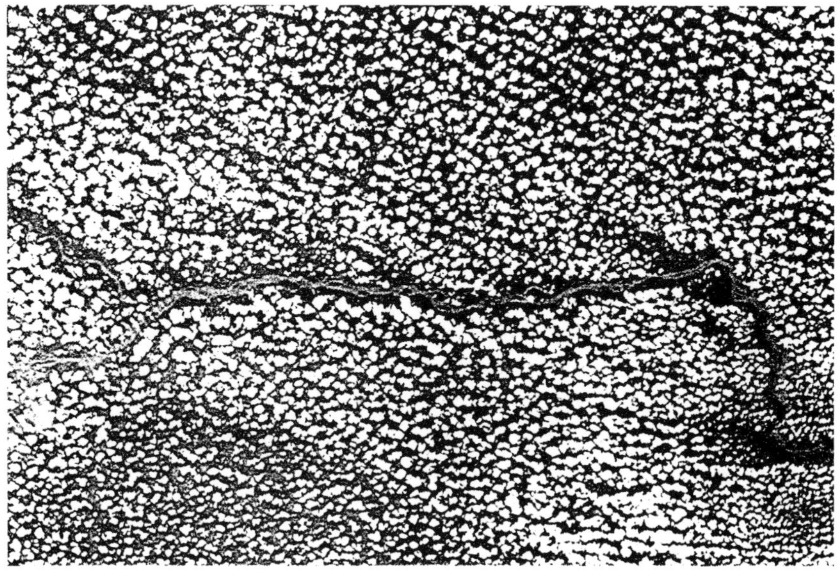

Figure 5.21. The river-breeze circulation caused by the Orinoco River is displayed by its effect on the otherwise uniform pattern of cumulus clouds above the jungle.

The regular pattern of clouds is interrupted over the rivers by the river-breeze circulation. The descending air above the rivers is shown by the absence of clouds, and the rising air is shown by slightly increased cumulus size at the clear boundaries.

The cloud-free zone shows that the river-breeze circulation is four or five times the width of the actual river.

Some work with numerical models has been done on the land and rivers breezes at Chongqing (Chungking) in China (Yan & Huang, 1988). Chongqing is a large industrial centre in south-west China where material from Sichuan Province is transported down the Yangtze River to the sea, 1500 km distant. Another large river, the Jialing, joins the Yangtze at Chongqing and the complex changes in wind systems must affect the air pollution and also the formation of fog. A numerical study which attempted to model the land–river breezes also considered the probable future changes to be expected after the Yangtze Gorges water conservancy project is completed, suggesting a 30% increase in the wind strength in some parts of the cycle.

Some of the results of the land–river breeze after the completion of the Yangtze Gorges project are shown in figure 5.22. Unfortunately, the present observational data are very sparse, making it difficult to check the truth of the model results.

As in the observed case of the Orinoco River, the width of the main circulation is several times the width of the actual rivers.

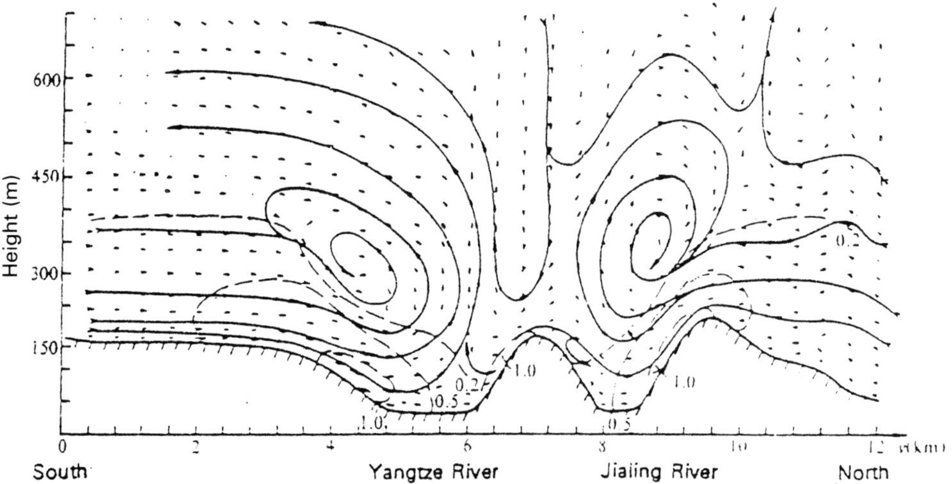

Figure 5.22. Circulation of the land and river breezes over Chongqing after the completion of the Yangtze Gorges water conservancy project. (After Yan and Huang, 1988.)

5.35 Local winds at urban heat islands

Towns have a generally higher temperature than the surrounding rural areas. The warming depends on a number of factors, among which are the amount of cloud and wind velocity. The difference is usually greatest at night, when 'islands' of warm air exist within major built-up areas. As a result, thermally induced local winds, similar to sea breezes, can be measured around the margins of large cities.

Measurements made across London's heat island (Chandler, 1961) show that sharp temperature changes exist at its boundary. Figure 5.23 shows a surface temperature traverse along the Lea valley across the northern boundary of London's heat island on a calm night in June. The trace shows a cliff-like margin of the heat island, in which a temperature change of 4 °C occurs in little more than 3 km.

Using timed drift of soap-bubbles, cool pulsating winds have been measured with strengths up to 1.5 m s^{-1}. Results such as these show how these light winds can have an effect on the form of the heat island. The cooler country-air enters the fringes of the city in a series of pulses, like miniature cold fronts, first on the leeward side than then on the windward side. Such local winds must also influence the frequency of fog formation and the distribution of pollution.

Surface-breeze effects have been measured at other cities as well as London but no simple relationship exists between the size of a city and the intensity of its heat island (Lee, 1992). The thermal effects of quite small towns are often surprisingly large, for example most of the features described for London have also been measured in the much smaller town of Leicester.

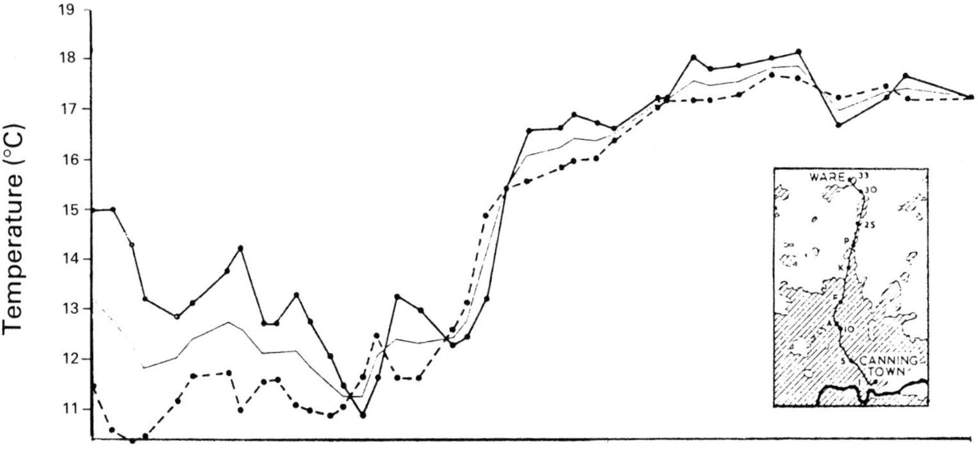

Figure 5.23. Lea Valley temperature traverse at night, 3–4 June, 1959, showing the sharp margin of London's heat island. (After Chandler, 1961.)

5.36 *Eclipse winds*

During a total eclipse of the sun the average temperature of the Earth's surface is reduced along a narrow shaded band. The temperature differences usually produce local winds similar to land–sea breezes. Eclipse meteorology is unique; the conditions imposed momentarily on the atmosphere by eclipses cannot be produced by any other events (Barlow, 1927).

The relatively sudden change in the radiation falling on the Earth during a total eclipse results in unusual changes in atmospheric motion. At a given place a total eclipse in all its phases will last about two hours, with a brief central period of complete absence of solar radiation. Figure 5.24 is an example of the changes in solar flux (radiation from the sun) during a day of total eclipse from

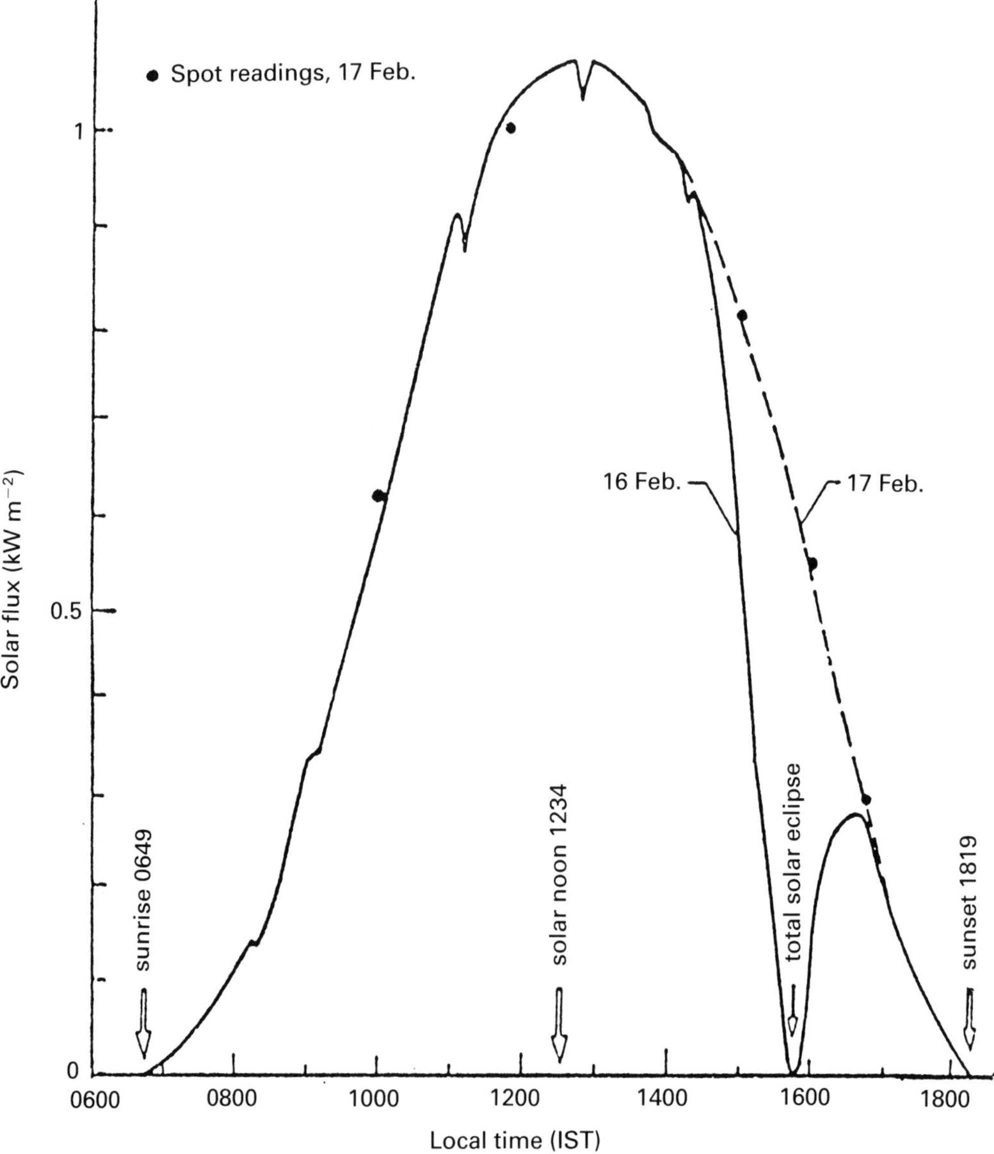

Figure 5.24. The changes of heat from the sun during a total eclipse of the sun at Raichur, India, on 16 February, 1980. (After Narashimha *et al.*, 1982.)

Raichur in India, when extensive measurements were made (Narashimha *et al.*, 1982). Measurements from this graph show that the rate at which the radiation was switched off during the eclipse was three times higher than during a normal sunset.

In the band of totality the temperature may fall as much as 3 °C, and some temperature reduction can also be detected as far as the 40% eclipse region. In the example shown in figure 5.24 the temperature at the end of the eclipse was 3 °C lower than usual for the time of day.

One of the secondary effects due to the temperature drop is a pressure change in the totality region. This averages a dip of 0.3 mb with the minimum about 20 minutes after totality. A very slight hump of less than 0.1 mb occurs at totality. These are typical and are of the order to be expected theoretically. The minimum convection and of pressure should come a few minutes later than the minimum temperature. The characteristics of wind changes observed are a decrease in velocity and gustiness, and also a shift in direction.

Winds, regarded as analogous to land and sea breezes, of up to 3 m s^{-1} were observed in the American eclipse of 8 June 1918 by Ramball & Ferguson (1919), but due to the comparatively small temperature gradient produced by the shadow the eclipse wind is usually small. However, there is evidence of wind maxima occurring about half an hour before and after totality. A short sharp gust was a marked feature at nearly all 154 Indian recording stations in the eclipse of 22 January 1898 (Eliot, 1900). These gusts sometimes reached 18 m s^{-1}, with an average wind of about 3.5 m s^{-1}. Was this a form of eclipse-breeze front?

6

Air quality

6.1 Pollution

It has been the custom to pour waste material, often poisonous, into the sea and the atmosphere. Now it is becoming clear that the ocean cannot be regarded as an inexhaustible sink for such material, and that the same also applies to the pollutants which we continue to pour into the atmosphere.

In the ocean we find that chemical and radioactive materials may remain concentrated and return to their source area. The same can also apply in the atmosphere, especially in the flow set up by the land- and sea-breeze circulation. The sea breeze limits vertical mixing since it is a relatively cool layer close to the surface and is therefore stable. In some cases the pollution is returned to the place from which it started, made more dangerous by chemical changes.

6.2 Pollution in the sea breeze

When a temperature inversion exists in the atmosphere over any large city, pollution problems are intensified because the stable layer acts as a lid to the dispersion of smoke and other airborne pollutants.

Pollution is a serious problem in the city of Los Angeles, which is illustrated in figure 6.1 in three different atmospheric situations. In figure 6.1(*a*) downtown Los Angeles is seen clearly. On this sunny convective day, smoke and haze are being carried upward and mixed into the surroundings, and as a result the visibility is very good. Figure 6.1(*b*) shows a light smog layer, which is shallow enough for the upper boundary to be visible. Figure 6.1(*c*) shows a day of severe smog, which is primarily of hydrocarbons and oxides of nitrogen from vehicle exhausts. The temperature inversion and stability of the atmosphere over Los Angeles is largely caused by subsidence from the Pacific anticyclone typically situated off the California coast.

Figure 6.1. Los Angeles in three different smog conditions. (*a*) Downtown Los Angeles on a clear day. (*b*) Smog building up in the inversion layer. (*c*) A day of severe smog. (Photograph by W. McDermott, 1961.)

Each day, when the sea breeze arrives, its refreshing effects are welcomed by the people living near the coast. However, it spreads the polluted air from Los Angeles to many places where it is strikingly evident in inland areas of Southern California. For example, measurements have been made at Riverside, about 60 km east of Los Angeles. Here, under typical conditions, as the air passes inland it is warmed enough to prevent cloud formation and plenty of sunshine is available to promote photochemical reactions in the heavily contaminated air.

6.21 Sea-breeze fumigation

During an onshore wind, such as occurs with a sea breeze, the cool, stable marine air flowing in the internal boundary layer becomes heated from below and may reach a neutral or superadiabatic lapse rate in the lower levels. With increased time and distance from the coast this heated zone continues to extend the mixed layer. The effluent from a tall chimney stack discharging into the stable layer disperses very little as it moves downwind. When it reaches the point downwind where the mixing layer extends upwards the material in the plume mixes rapidly downwards to cause 'fumigation'. High concentrations of the effluent then reach ground level.

The process of fumigation is illustrated in figure 6.2, which shows the contrast between the effects of a tall stack and one which is already beneath the internal boundary level. These measurements were made near the east coast of Lake Michigan and show the concentrations of oxides of sulphur measured from an aircraft and also on the ground. The plume from the higher chimney does not mix much until it comes in contact with the mixed layer at a distance of 5 km from the coast.

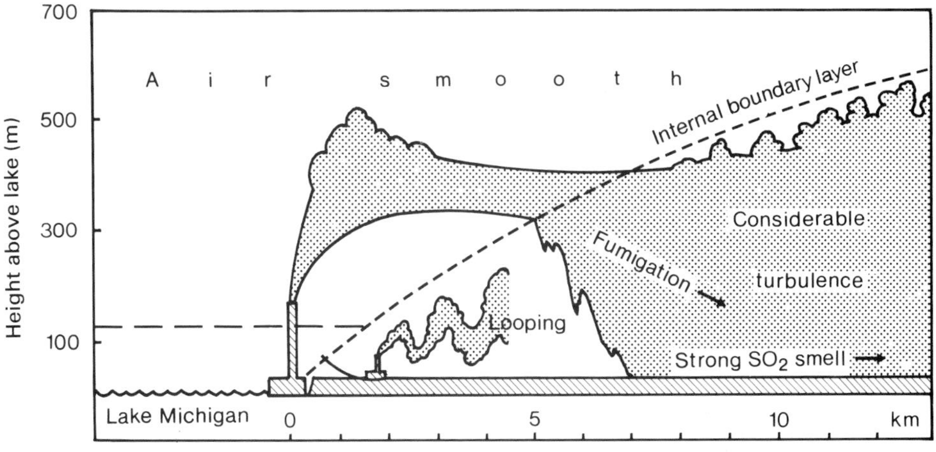

Figure 6.2. Plumes from a tall and a low chimney stack near the east coast of Lake Michigan showing the process of fumigation observed as the plume from a high chimney moves moves from stable air into the mixed layer above the land. The concentrations of oxides of sulphur were measured by a spotter aircraft around 1400 CST, on 25 June 1970.

6.3 Pollution at sea-breeze fronts

Perhaps a dozen times a year the demarcation between the polluted marine air at Los Angeles and the clean desert air becomes very sharp. An example of one such boundary is shown in figure 6.3, (Stevens, 1975). Scaling the photograph showed the smog bank to be 1000 m thick. At the passage of this front, marked changes in temperature and humidity were observed and also changes of oxidant and peroxyacetyl nitrate (PAN), which both increased abruptly, as shown in figure 6.4. There was an increase in moisture, but the relative humidity was only 30%, showing that the visible part of the smog was not fog, but was the aerosol. The trace of wind direction shows clearly the change at 1400 h from the earlier regime of light wind, with passing patches of convection, into a stronger and much steadier flow in the sea-breeze layer behind the front.

This kind of pollution build-up at a front appeared in the illustration of figure 1.2 in Chapter 1, showing the well-known 'Middlesborough Muck' in

Figure 6.3. Sea-breeze front polluted by photochemical smog at Riverside, California, in the early afternoon of 16 March 1972. (*a*) Wind strength and direction, (*b*) wet and dry bulb temperature, (*c*) values of oxidant and peroxyacetyl nitrate (PAN). (Photo courtesy of G.R. Stevens.)

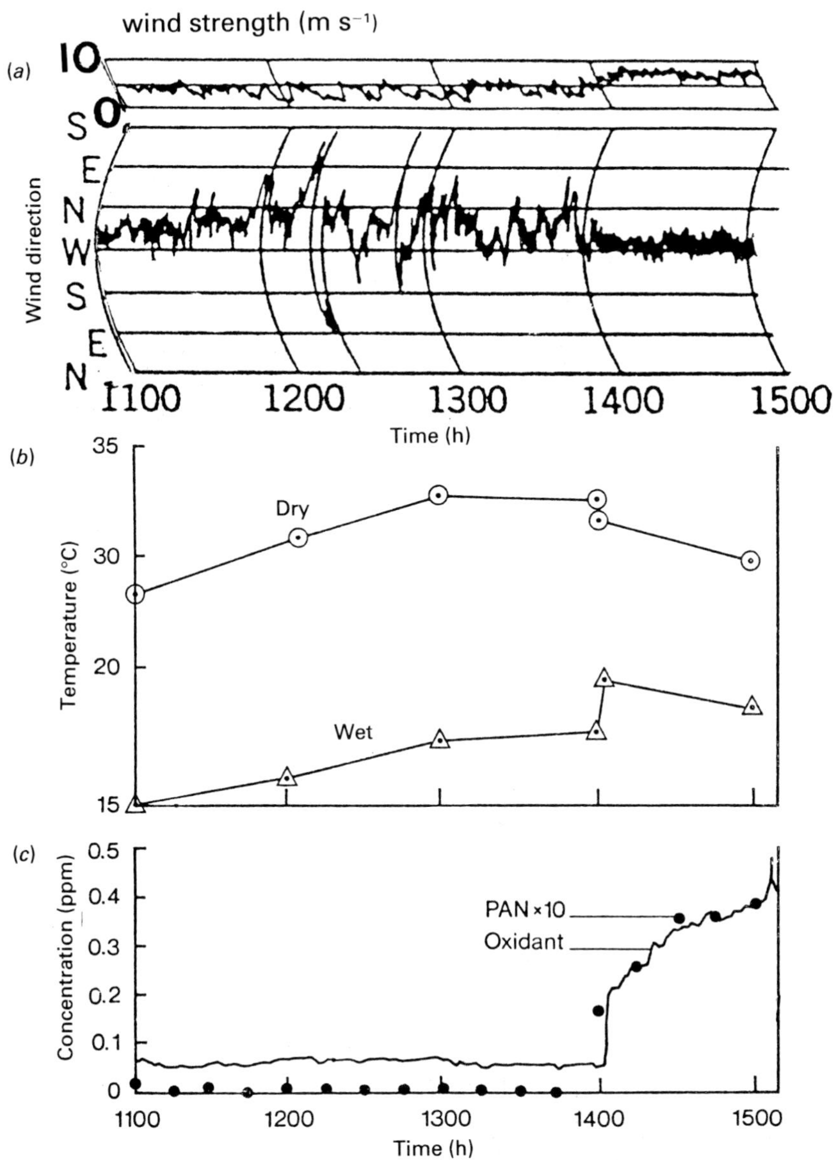

Figure 6.4. Changes at the passage of the sea-breeze front at Riverside, California, on 16 March 1972. Variables are wet and dry bulb temperature, oxidant and peroxyacetyl nitrate (PAN). (After Stevens, 1975.)

the sea-breeze front advancing from the smoky town of Middlesborough in north-east England. This phenomenon was well-known to the forecasters at the nearby RAF Station at Thornaby-on-Tees where it commonly reduced visibility from 17 km to 800 m within 5 minutes during the summer months, as the

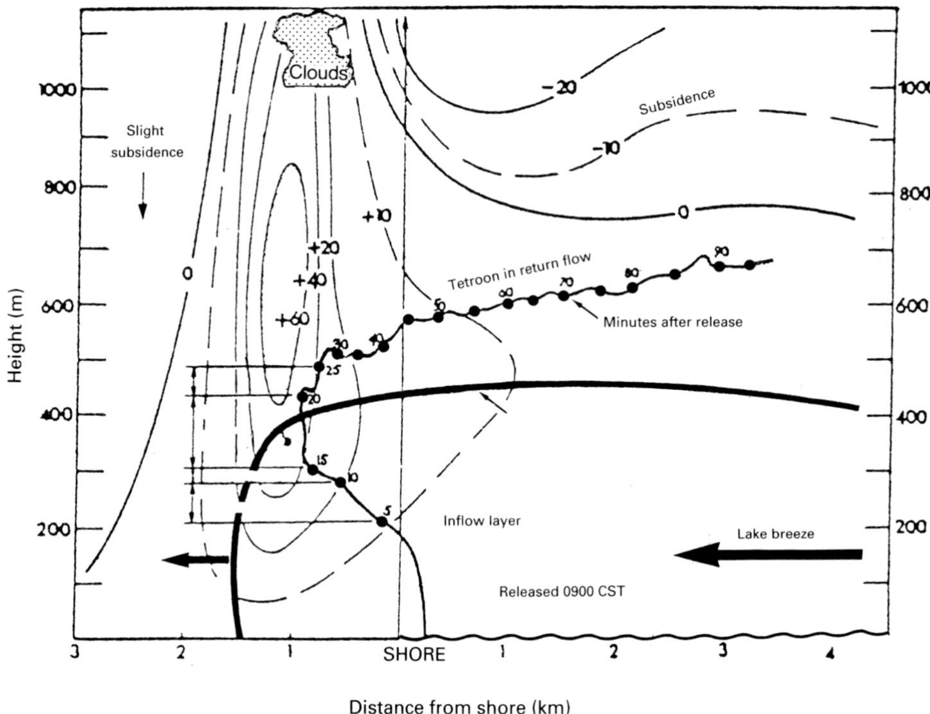

Figure 6.5. Vertical air movement near the front of a lake breeze at Chicago on 13 August 1972. The constant volume balloon (tetroon) which passed through the head of the front later reversed its direction. The tetroon's position appears every 5 minutes and the mean 5 minute vertical motions are also plotted. The vertical line with the wind-arrows is the shoreline pilot balloon wind sounding at 0845 CST (After Lyons & Olsson, 1972.)

sea-breeze front carried household smoke and chemical pollutants from the industrial plant at Middlesborough.

In addition to household and industrial pollution the main causes of emission responsible for this kind of pollution in the sea breeze are vehicle exhausts. This source is one of those responsible for the formation of photochemical smog produced by chemical changes caused by sunlight. This is a serious problem in many modern cities, such as Tokyo, Athens, Mexico City and many others.

6.31 Trajectories near sea-breeze fronts

A good way of determining trajectories in the neighbourhood of a sea-breeze front is the use of constant-volume super-pressurised balloons. A popular type of such balloon, used to obtain Lagrangian air motions, is the tetrahedron-shaped 'tetroon' with a volume of about one cubic metre. The high drag coefficient (0.8) and the low expansion rate (up to about 4%) allow the balloon to respond to vertical motions of the air.

Many tetroon studies have been conducted using a radar-transponder system to trace the balloon's position and aeroplanes or helicopters have also been used to track the balloon visually. They were used on the coast of California in the 1960s (Angell & Pack, 1965) and also extensively in the Chicago district, in studies of pollution transport in the lake breeze (Lyons & Olsson, 1972).

The fate of pollution once it meets the sea breeze can be understood using the remarkable similarities between atmospheric field observations and well-documented gravity currents in the laboratory. Ambient and sea-breeze air mix mainly at the rear of the elevated head of the front, mostly at about 1 to 2 km from the leading edge and it is this mixed air which constitutes the main return flow above the incoming sea-air. Hence, the pollutants will ascend at the sea-breeze front and the only way in which they can enter the oncoming sea-air is if they are convectively mixed down from the return flow. Figure 6.5 gives a tetroon trace showing an example of this behaviour.

A summary is shown in figure 6.6 of the observations from the Chicago workers from tetroon and aircraft measurements. The 'wall of smoke' behind the front due to fumigation is a striking feature and some recirculations observed from the tetroon are included.

6.4 Diurnal recycling of pollution

Some recirculation of sea-breeze polluted material exists not very far behind the sea-breeze front; however, the really serious problems of pollution carried by the sea breeze are related to the land- and sea-breeze reversal. This gives a mechanism for a complete layer of polluted air to be maintained at high concentration and returned to the same locality 24 hours later.

Tetroon trajectories originating near a coastline confirm this behaviour. Figure 6.7 shows the trajectories of several tetroons released from a point on the California coast near Los Angeles (Pack & Angell, 1963). These show, in one case, a passage as far as 27 km out to sea, followed by a return in the sea breeze. Another travelled 30 km to the west along the coast, and a third one, released late in the day, moved almost straight inland. Three short traces, made at a point 30 km along the coast, show a hint of sea breeze veering with time.

6.41 Air pollution at Athens

Athens, a city of over 3.5 million inhabitants, is located in a small natural basin formed by mountain ranges on three sides and the Gulf of Saronikos to the

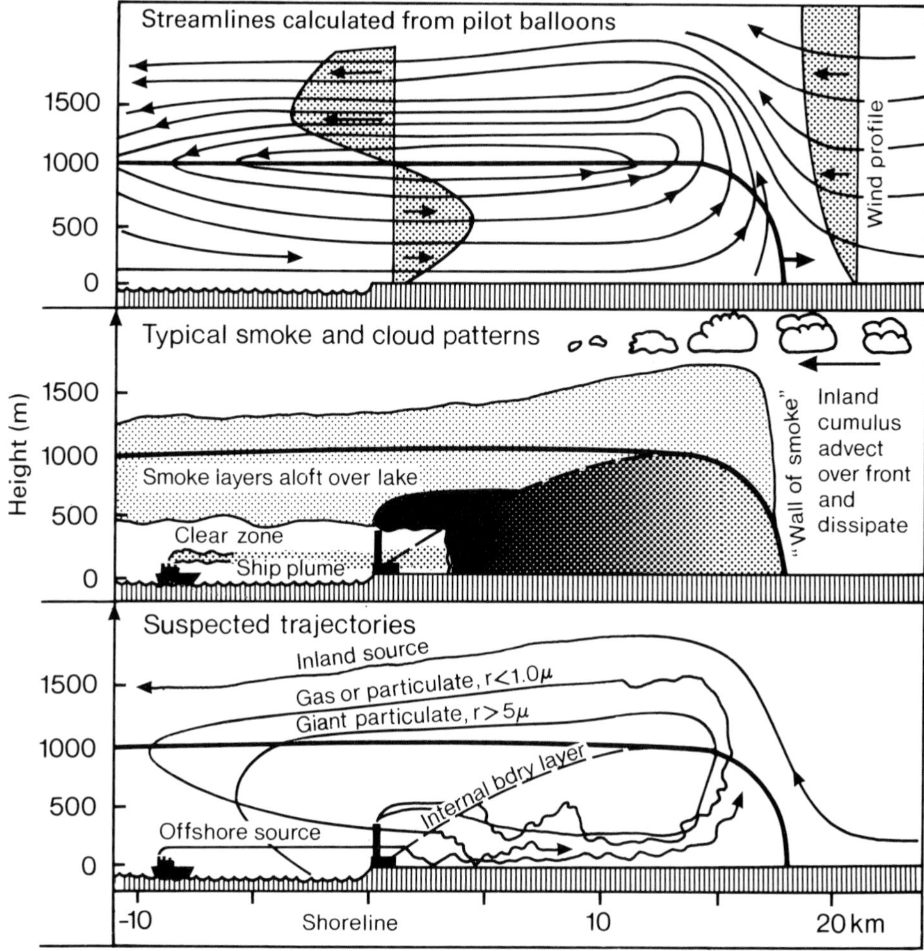

Figure 6.6. The form of the lake-breeze front near Chicago, measured using tetroons and aircraft. Special features are the 'wall of smoke' and the evidence for recirculation. (After Lyons & Olsson, 1972.)

south-west; see figure 6.8. In this small basin are concentrated 50% of all automobiles registered in Greece and 40% of the Greek industrial factories.

The effect of sea breezes on the photochemical smog level at Athens has been studied in detail. It has been made clear that nearly all photochemical smog episodes with values exceeding the US Air Quality Standard of 120 ppb ozone were accompanied by well-developed sea breezes (Gusten & Heinrich, 1989).

Due to local air circulation in the closed topography of the Greater Athens basin, precursors of ozone and PAN (i.e. NO_x and hydrocarbons) are transported to and accumulate over the Gulf of Saronikos during early morning

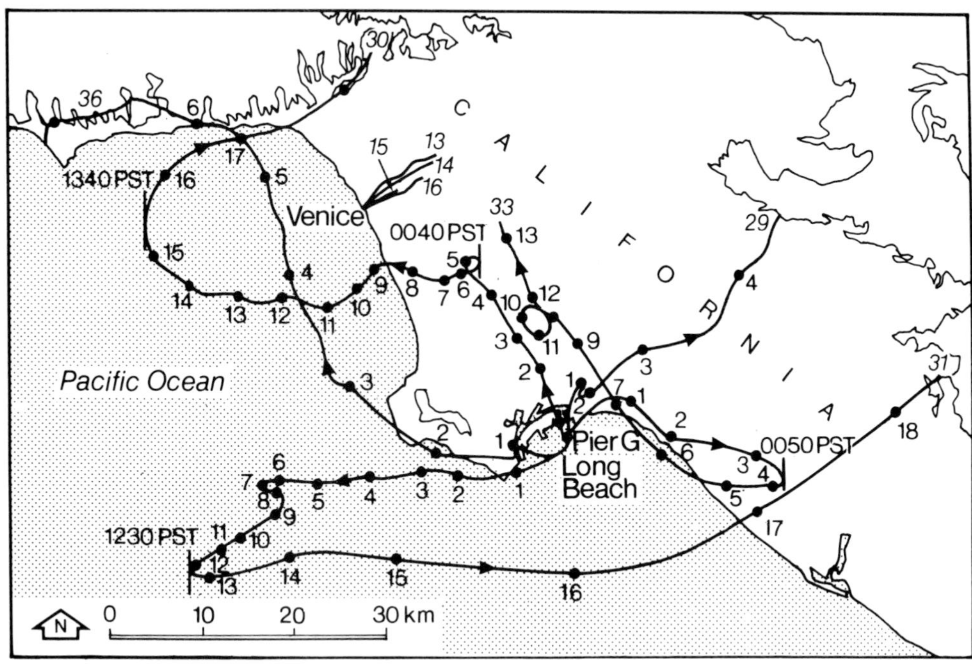

Figure 6.7. Reversal in the sea-breeze circulation, shown by a series of tetroon tracks from Long Beach, California. In one case a passage in the early morning as far as 27 km out to sea was followed by a return to the land in the sea breeze. (After Pack & Angell, 1963.)

while the land breeze is blowing. Around noon, when the sea breeze sets in, the photochemically produced ozone and PAN formed over the sea are brought back to the coast and on to central Athens, where they increase the local concentrations by a factor of 3 to 5.2 within a few hours (figure 6.9).

6.42 Sea breeze and mountain winds

The sea breeze and winds towards the mountains often combine when high land is near the coast, and the reverse flow down the mountains at night is often more significant than any land breeze which might otherwise occur.

A channelling effect observed when the sea breeze passes through a mountain pass in Japan is shown in figure 6.10. It is clear that the spread in the mountain area is much restricted (Ueda, Mitsumoto & Kurita, 1988).

Examples of the reversed advection of pollution in the sea breeze are given in figure 6.11. This schematic diagram is based on field work carried out in Japan (Kurita, Ueda & Mitsumoto, 1990) and shows the effect of the land and

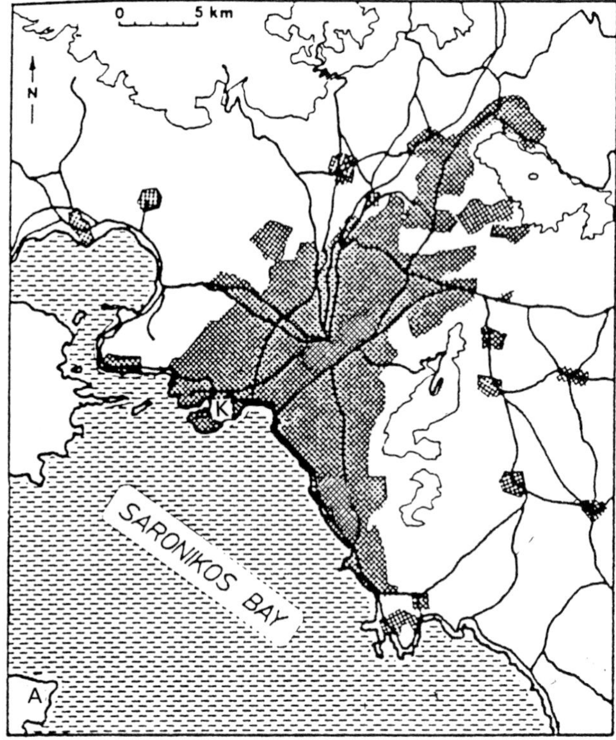

Figure 6.8. Greater Athens and Saronikos Bay, showing ozone monitoring stations at the Island of Aegina (A) and Kastella (K) on the shoreline. (After Gusten & Heinrich, 1989.)

sea breezes combined with mountain winds during one 24-hour cycle. One striking feature is the return of the previous day's pollution after it has been carried out to sea. Another feature shown is the division of the pollution which has been carried inland on the sea breeze. This divides at the top of the pass, some it returning in the descending mountain wind, while the rest travels across and descends towards the opposite coast.

6.5 Chemistry of sea-breeze pollution

Many large cities are built near the coast and their pollutants issuing into the atmosphere are therefore subject to the land- and sea-breeze circulation system. When there is high ground near the coast the total circulation system must also include the effects of the mountain and valley winds.

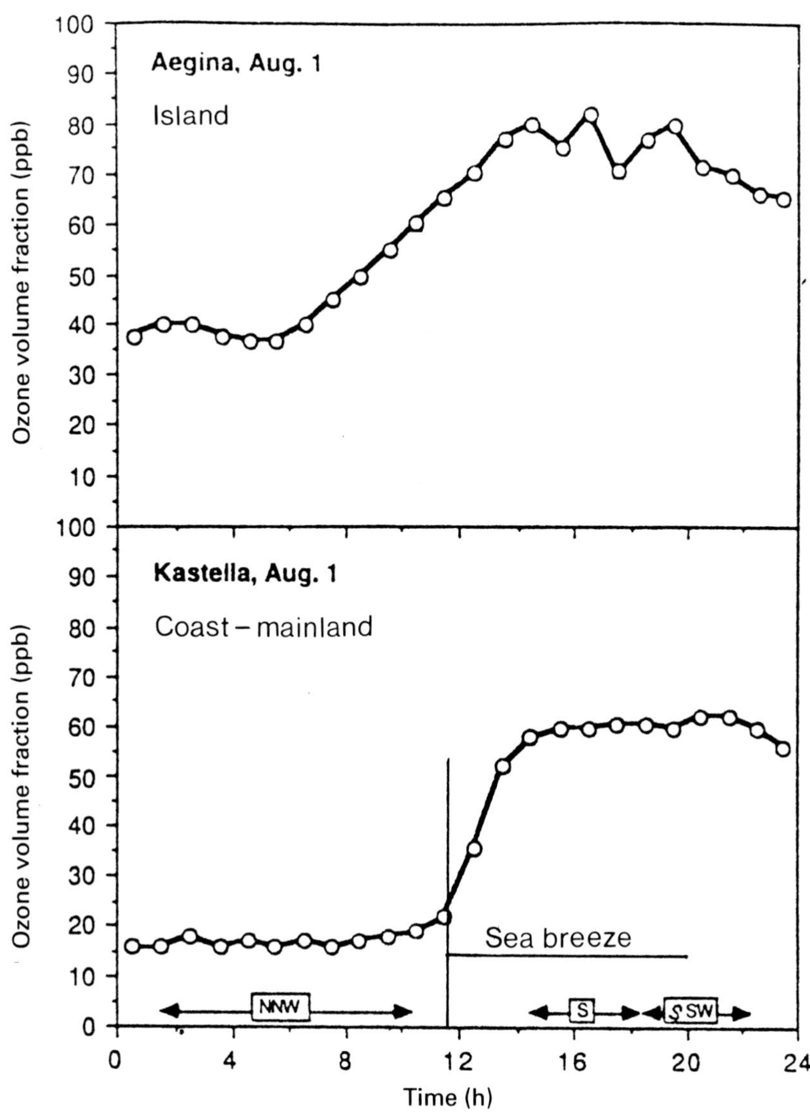

Figure 6.9. Ozone concentrations at Aegina and Kastella on 1 August 1984, showing a rapid increase with the sea-breeze onset at the shore. (After Gusten & Heinrich, 1989.)

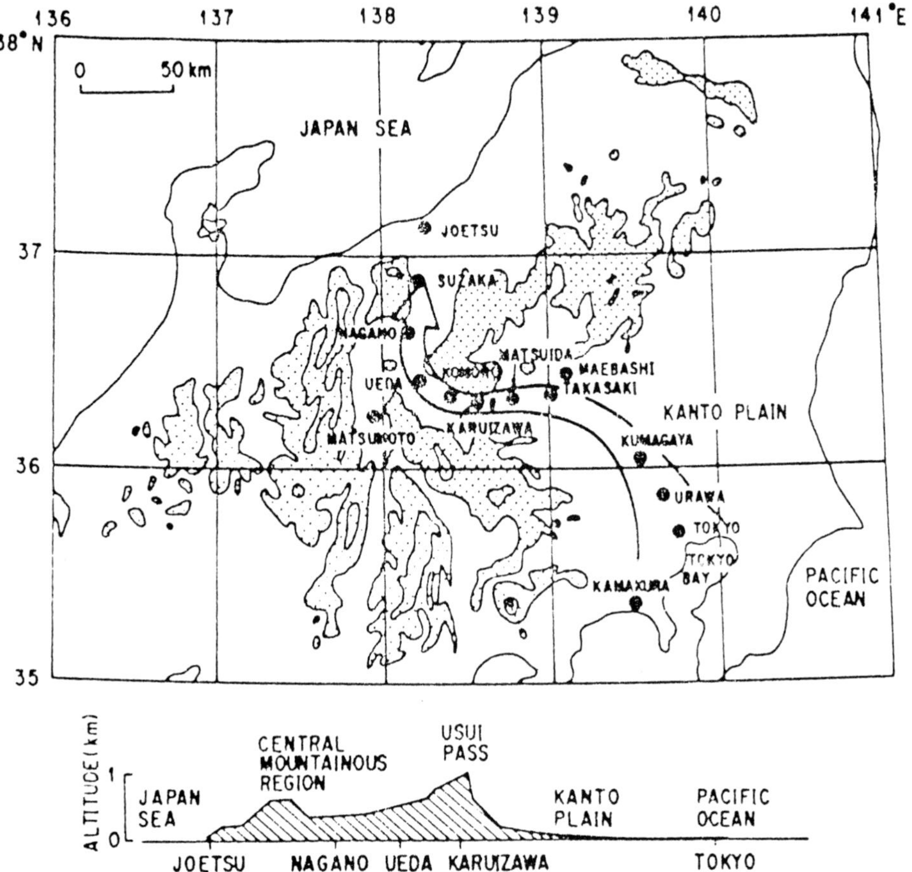

Figure 6.10. Map of the mountain area near Tokyo with observation sites. Regions of altitude greater than 1000 m are hatched. The path of pollution transport is roughly denoted by the curved arrow. The vertical section along the path is show below. (Courtesy of H. Ueda.)

6.51 Chemical sources

The air pollutants involved can be divided into the following four classes, according to their chemical composition.

1. NITROGEN-CONTAINING COMPOUNDS

These are oxides of nitrogen, NO_x, from combustion sources including power plants and motor vehicles. The irradiation of a system containing these nitrogen compounds and air results in the oxidation of NO after the production of ozone. However, the addition of some of several different organic compounds greatly accelerates the photo-oxidant processes.

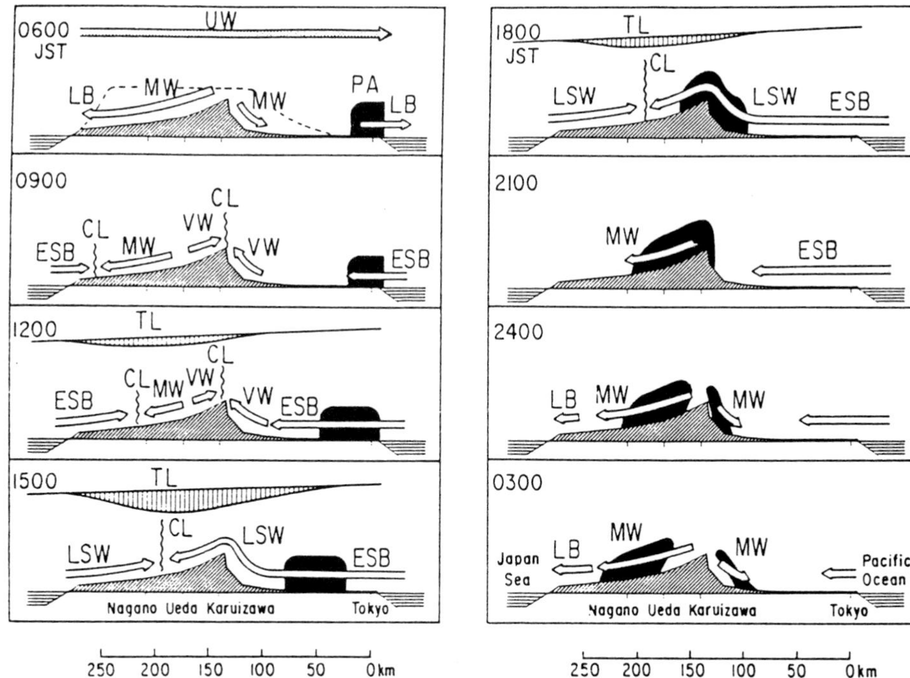

Figure 6.11. Schematic diagram of transport processes of air pollution from the coastal region to the inland mountainous region in Japan, which reaches a height of more than 1000 m. CL, convergence line; PA, polluted air mass; SB, sea breeze; EB, extended sea breeze; LB, land breeze; MW, mountain wind; UW, upper wind; VW, valley wind; TL, thermal low; LSW, large scale wind blowing towards thermal low. (Courtesy of H. Ueda.)

2. CARBON-CONTAINING COMPOUNDS: REACTIVE HYDROCARBONS

Many hydrocarbons, including aldehydes, ketones and unsaturated hydrocarbons are very reactive in the atmosphere and are collectively known as RHC, reactive hydrocarbons. They occur from combustion of materials and also to a smaller degree from solvent use. They enter the atmosphere in various ways, of which the greater source is emission from motor vehicles, which contributes 86% of the reactive oxidants found in Los Angeles (Leighton, 1961).

3. SULPHUR COMPOUNDS

These are SO_2 and H_2S. The more important is SO_2, which oxidises with water in the atmosphere, finally producing H_2SO_4, or suphuric acid, which is advected in the main synoptic wind systems as a constituent of acid rain.

4. AEROSOLS

Atmospheric reaction processes are frequently accompanied by aerosol emission. These are derived from motor vehicles and industrial emissions, together with

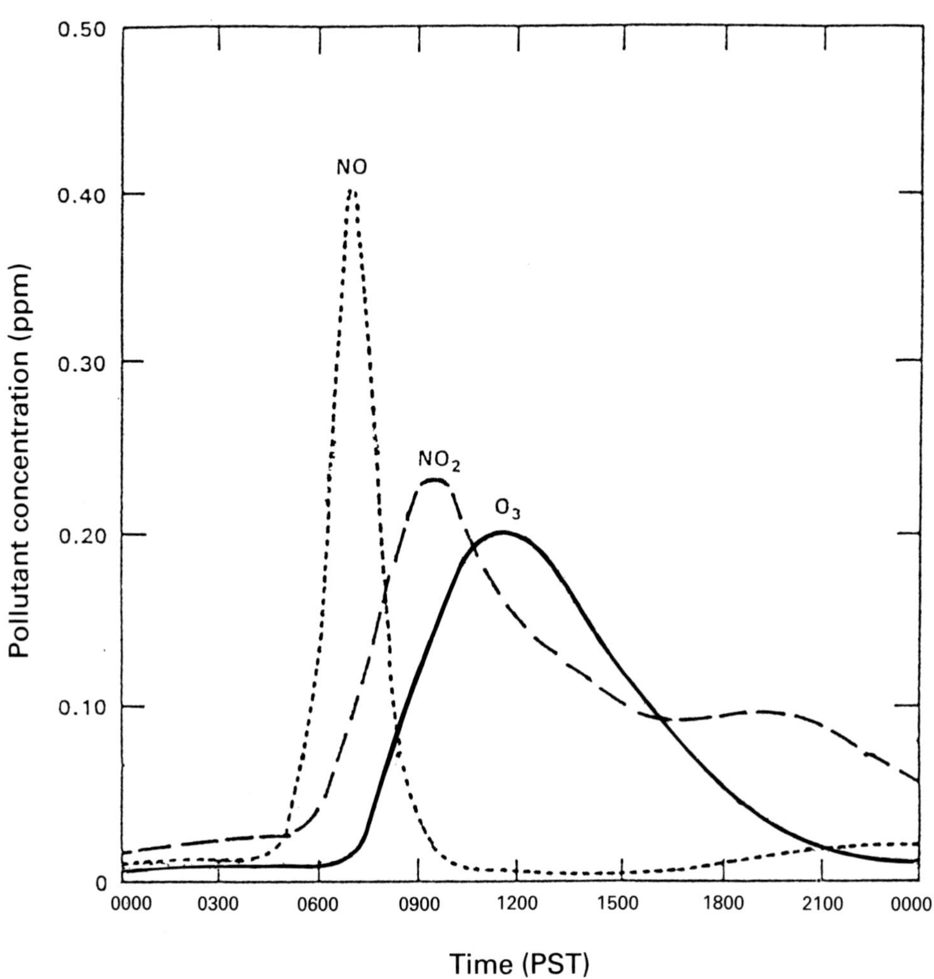

Figure 6.12. Average diurnal variation in concentrations of selected pollutants at Los Angeles, July 19, 1965. (From US Department of Health, Education and Welfare.)

some natural aerosol production. The latter consists mainly of dust raised by the wind, with some sea salt, but motor vehicles again form an important source.

6.52 Transport and mixing

The materials are carried inland during the early part of the day and this flow may soon be extended near any mountains, where the flow extends upwards in a layer above the slopes. Later in the day the land breeze combines with valley winds to carry the polluted air back towards and above the ocean.

6.53 Transformation and photochemical smog

The chemistry of ozone in the polluted lower atmosphere has been extensively studied for the past 30 years. The rate of ozone formation varies in a complex way which is dependent on the ratios and concentrations of the reactive hydrocarbons and the nitrogen compounds.

These gases enter the atmosphere in various ways: the largest contribution is generally from motor vehicle exhaust, which contains unburned and partially burned petrol. For example, about 66% of the total organics and 86% of the reactive organics found in the Los Angeles atmosphere are from motor vehicles (Seinfeld, 1975).

Typical observations made in a polluted atmosphere, which show the results of emissions into the atmosphere and the consequent transformations, are shown in figure 6.12. The diurnal pattern begins with the emission of carbon monoxide and NO_x and hydrocarbons, which act as accelerators. Ozone accumulation begins when most of the NO has been oxidised. The ozone concentration reaches a maximum at around 1200 h and then declines. The increase is due to the additional emissions of NO_x during the day, and the decrease is due to dilution in the atmosphere, and by interaction with the ground surface, which is a major sink for atmospheric ozone.

We can summarise the diurnal pattern as follows:

Morning: NO_x, RHC are emitted in the morning rush hour.

Afternoon: O_3 formation increases as solar insolation increases.

Evening: NO_x is emitted again during the evening rush hour.

Night: Solar production of O_3 ceases; its concentration then decreases due to mixing and ground absorption.

7

Sea breeze interactions

7.1 Sea-breeze convergence zones

The sea breeze usually starts to move inland at right angles to the land boundary; when the coast is straight a more-or-less uniform sea breeze spreads inland. If the coast is not straight, then the flow inland will not be uniform, but will converge or diverge according to the curvature of the coastline.

Figure 7.1 is a simple diagram showing how on a curved coastline the sea breeze converges at convex coasts and diverges when the coast is concave. As most coastlines are curved in some way, this effect is very important in the development of the sea breeze, and areas of convergence and divergence are frequently formed. Incoming air which must rise at the convergence line is often associated with convective clouds and rain.

7.11 Shear lines

Wind shear is the rate of change in wind velocity (speed and direction) in the direction perpendicular to it; this may occur over an area but when it occurs along a line this is called a shear line. Many convergence zones form shear lines, of which a well-known example is the sea-breeze front.

7.2 Effects of headlands and peninsulas

Convergence zones are especially common at headlands and peninsulas, both areas of strongly convex coastlines. Their existence may be deduced from the formation of a line of clouds in an otherwise clear sky. Such lines can often be spotted when travelling by aeroplane. An example is given in figure 7.2, taken

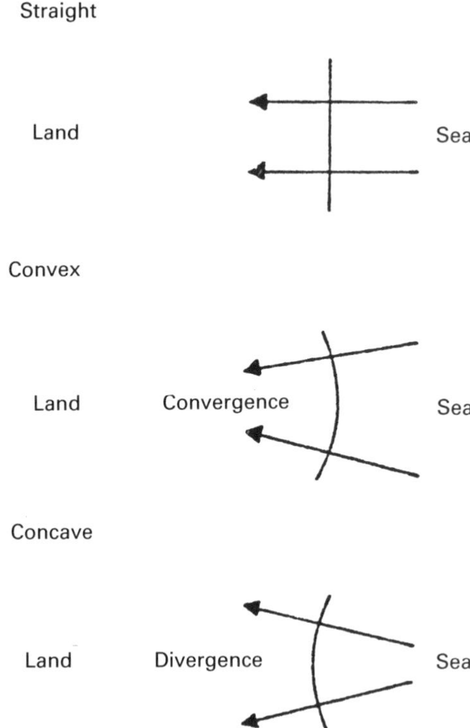

Figure 7.1. The effect of curved coastlines on the strength of the sea breeze. It is strengthened by convex coasts and weakened by concave ones.

from a satellite on 8 July 1981, which shows a line of cloud in the centre of Denmark. The winds were light, and the clouds have formed in the area of convergence formed by the sea breezes on the two sides of the country.

Convergence effects are noticeable even on a peninsula as large as Florida. As far back as 1948 it was shown (by Byers & Rodebush) that the meeting of the two sea breezes from the east and west boundaries of Florida was responsible for the large incidence of thunderstorms there.

Satellite photographs often show large-scale cloud formations which have developed at sea-breeze convergence zones and another example is shown in figure 7.3. This shows the effect of the two peninsulas extending west towards Pembroke in South Wales and the larger peninsula of Devon and Cornwall, which often produce convergence zones as shown in this photograph. The convergence here produces a line of cumulus clouds which steadily grows wider and, if the wind is fairly fresh, extends far to the east. Although these lines start off due to sea-breeze convergence, they develop a circulation which develops far beyond the original peninsulas. From Cornwall the line sometimes extend

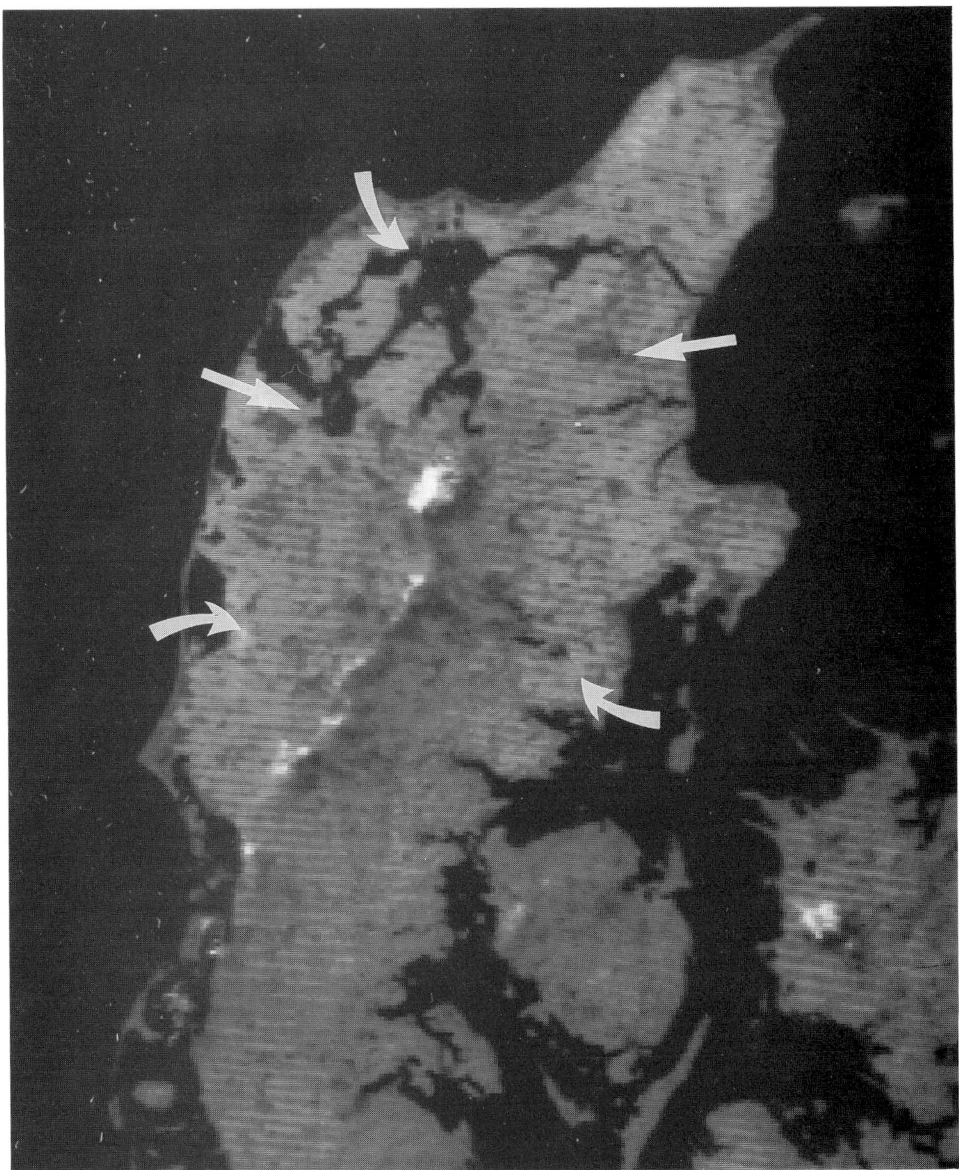

Figure 7.2. The line of clouds in the centre of Denmark is formed by convergent sea breezes from the opposite coasts. From satellite NOAA 6 on 8 July 1981. (Photo by University of Dundee.)

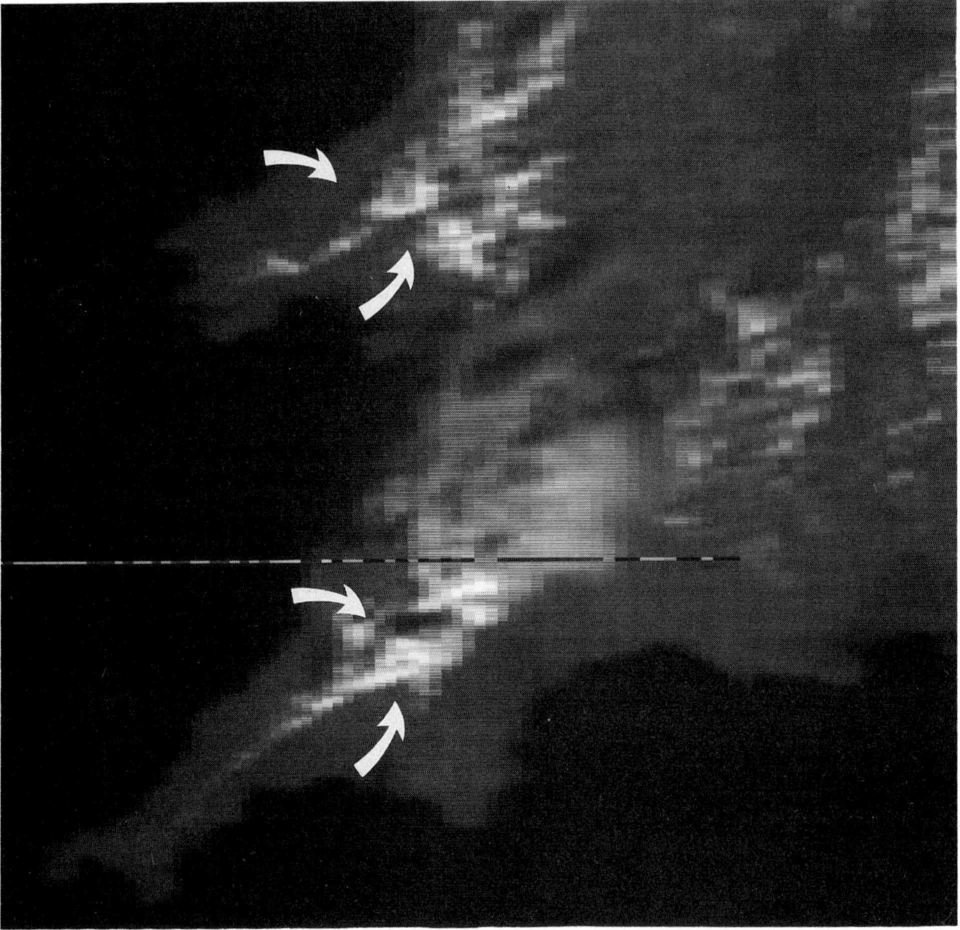

Figure 7.3. Sea breezes forming convergence zones in south-west England and Wales on 23 May 1989. Clouds show that the circulation has developed far beyond the original peninsulas. From satellite NOAA 11. (Photo by University of Dundee.)

as far as London and by the time it reaches the city it may have spread out to form several lines of cumulus (Bradbury, 1990).

A particular sea-breeze convergence zone which has been well documented (Findlater, 1971) is the one which forms in the Somali Republic near the promontory at the north-east corner of Africa. Measurements of the converging surface winds from the Bay of Aden and the Indian Ocean have led to maps such as that in figure 7.4, which gives the mean surface wind pattern in July. The convergence is shown lying along the range of mountains, some at heights above 1000 m. The airflow in this district has especial importance in the study of locust pests and their development and distribution.

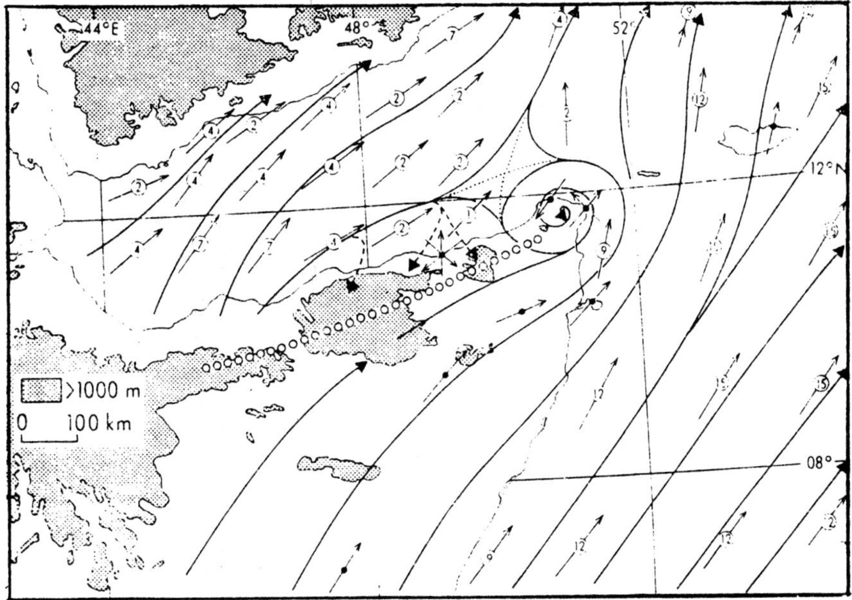

Figure 7.4. Convergence zone, shown by ooo, at the north-east corner of Africa in July. The convergence lies along the range of mountains, which are higher than 1000 m. (From *Meteorological Magazine*, by permission of HMSO.)

7.3 Bifurcations

An advancing sea breeze may be divided by a mountain obstacle into two separate streams which later come together again and form a convergence zone. Good examples occurs near Los Angeles, where two different mountain ranges divide the sea breeze into separate sections. These meet again further inland to form two separate convergence zones. Figure 7.5 shows how in the north the Santa Monica Mountains divide the flows, which meet again in the San Fernando Valley, and the way the Elsinore shear line derives from the Santa Ana Mountains in the south-east.

7.31 The San Fernando convergence zone

A detailed analysis made of air pollution trajectories at the shear line formed by the flow of sea-air round the Santa Monica Mountains is outlined in figure 7.5. It was found that the westerly branch of clear air met the heavily polluted south-east flow from Los Angeles in the western part of the San Fernando Valley (Edinger & Helvey, 1961). The resulting shear line progressed eastwards

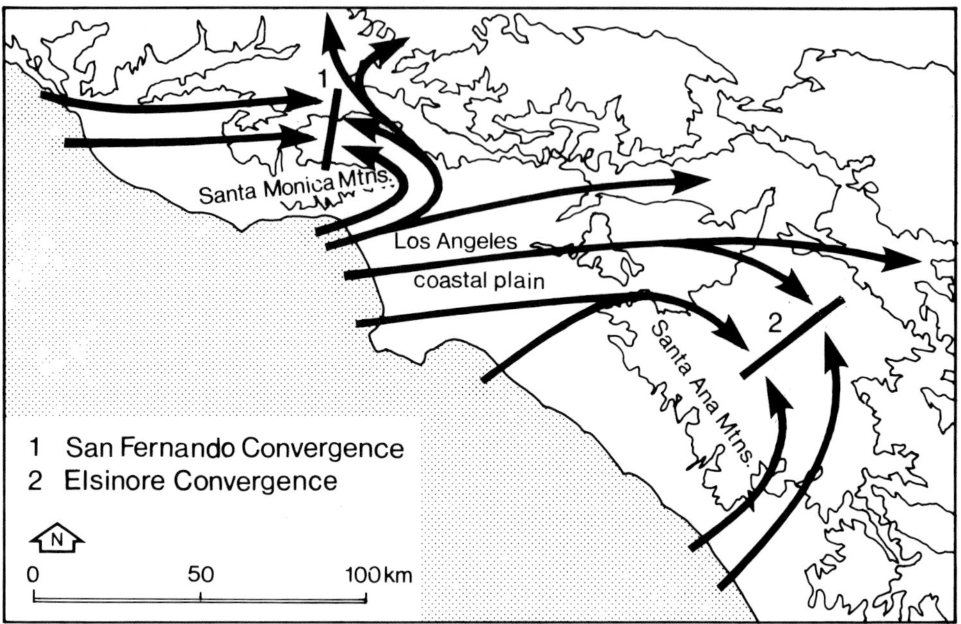

Figure 7.5. Typical sea breezes in the Los Angeles district, which flow round two mountainous areas and converge to the north and south. (After Edinger & Helvey, 1961.)

to the central part of the valley during the afternoon. Airborne measurements showed that much of the polluted air was forced aloft along the shear line, while some of it was diverted north into the desert.

7.32 The Elsinore shear line

Two currents flow around the Santa Ana Mountains, reaching Elsinore from both the north-west and the south (Aldrich, 1970). Identification of the airflow trajectories is aided by the polluted air from the Los Angeles Basin which makes a 'smog front' in the late afternoon. As the smog front develops near Elsinore the shear line only moves slowly, with the result that there is a strong ascent of polluted air. This line of 'lift' along the shear line has been exploited by glider pilots and more details are given in Chapter 9.

7.4 Meetings of fronts

Careful analysis of the unexpected appearance of cumulus or cumulonimbus sometimes shows this to be the result of the interaction of two already well-marked convergence lines. It has been found that two well-formed sea-breeze fronts may approach each other and create strong convergence at their meeting place; the development of two such convergence zones is described in the following two case studies.

7.41 Clashing fronts in south-east England

A light westerley airstream lay across southern England on 12 June 1963, with a minor trough from the Wash to Hampshire. By 0900 GMT a streamline analysis revealed two convergence lines, one at a weak trough and the other associated with the onset of the sea breeze from the south coast.

As the day progressed a second sea breeze moved inland, north of the Thames estuary, and a third one moved inland across the Wash. There was no low cloud in the areas on the coastal side of the sea-breeze convergence lines, but by 1300 h cumulus or cumulonimbus had become organised and an hour later thunderstorms were reported from these areas. As can be seen in figure 7.6, in each of the three areas where convection led to thunderstorms, a sea breeze closely approached the pre-existing convergence line.

It is noteworthy that all the analysis of surface charts in this investigation (Findlater, 1964), which led to the tracing of the convergence zones, was carried out using only the information available from the teleprinter broadcast from the Central Forecasting Office, Bracknell.

7.42 The 'sea-breeze bottleneck'

This name has been given by glider pilots to the narrow band of land in the south-west of England, north of Lyme Bay, which must be traversed during flights to the west. The author, who was convalescing from an illness in 1964 in this part of the country, was able to lie back and watch a double sea-breeze front event developing, on 22nd August 1964. Fortunately, a camera and recording thermo-hygrograph were available and many sightings were available on this day from glider pilots who were flying a task on an eastern course from Dunkeswell.

On this clear sunny day the sky was filled with small cumulus clouds and by midday a sea-breeze front in the south could be seen from South Petherton, clearly marked by trailing curtain clouds. A little later another sea-breeze front could be seen approaching from the north; this passed overhead and could be seen meeting the other front from the south. The map in figure 7.7 shows the situation which had developed by 1530 GMT.

Later in the day the front to the west of Sidmouth curved back to the north-east and a section of sea-air separated and moved east. Further to the east,

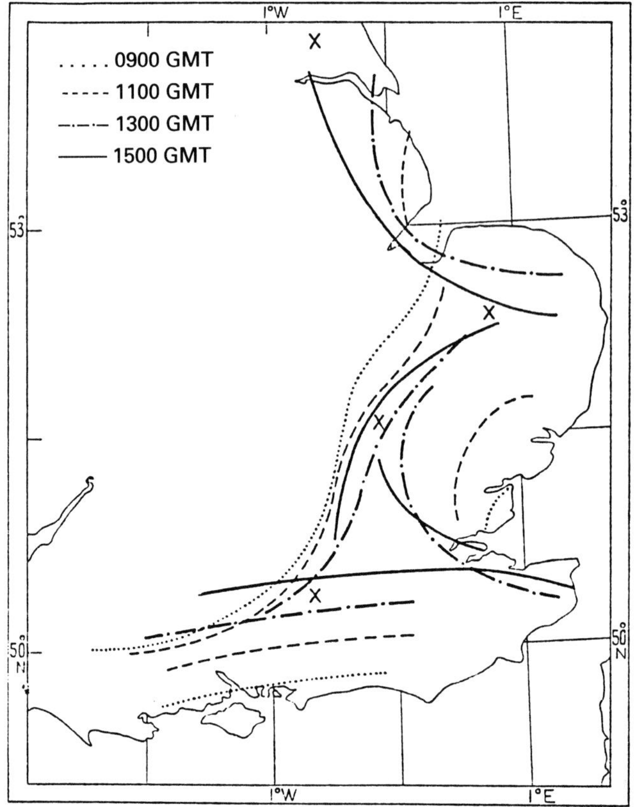

Figure 7.6. Three areas in south-east England where thunderstorms developed on 12 June 1963. In each of the places marked a sea-breeze was observed to approach a pre-existing convergence line. (From *Meteorological Magazine*, by permission of HMSO.)

beyond Bridport and Dorchester, a massive line of big cumulus was visible and this continued until late in the evening, when all other clouds had disappeared. An example of a similar situation is shown in figure 7.8, taken north of Bridport at 1700 GMT on 5 August 1964.

This meeting of two fronts in Somerset and Dorset is not unusual and confirmation comes from a satellite photograph in figure 7.9, taken on 8 July 1983, which shows two clear convergence lines in the same area.

7.5 Head-on collisions

Head-on collisions between two sea-breeze fronts or other similar meso-fronts have been observed. In one case two radar lines of echoes from fronts in south-

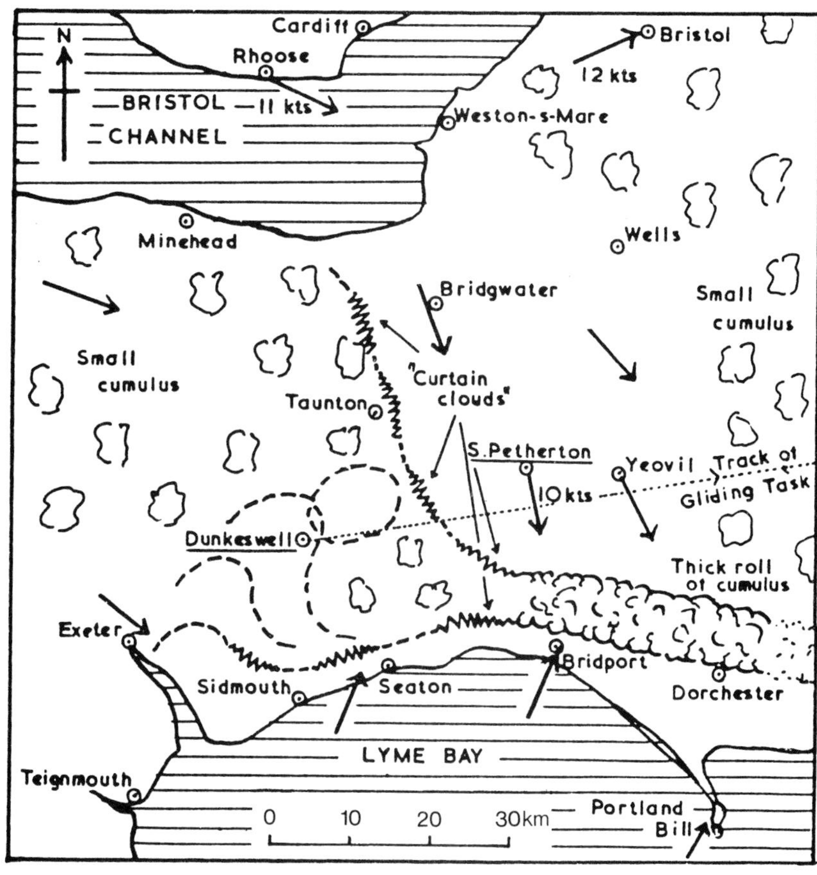

Figure 7.7. Two sea breezes converging in south-west England, north of Lyme Bay, at 1530 GMT on 22 August 1964. A thick roll of cumulus had formed north of Bridport and Dorchester.

ern Florida (Boyd, 1965) were seen to meet head-on and cross and then proceed independently, at first sight as if nothing had happened.

Head-on collisions are easier to understand than the 'glancing' collisions described above, because in these cases the flow patterns may remain roughly two-dimensioned. This makes it easier to investigate their behaviour using laboratory experiments and by mathematical modelling.

Figure 7.8. A thick roll of cumulus cloud north of Bridport formed in a sea-breeze convergence zone. Looking west at 1700 GMT on 5 August 1969.

7.51 The 'double coast' sea breeze

In a simple 'double coast' model (Clarke, 1984) sea breezes form on both sides of a peninsula with two parallel coastlines and approach each other during the day. One is assisted and the other resisted by the existing gradient wind. In time the former becomes more vigorous than the latter and eventually the two sea breezes collide, nearer one coast than the other.

The collision raises cool air in a sharp upwards bulge. When this bulge collapses, two bore-waves may propagate into the stratified layers formed by the sea breeze on both sides of the collision site. These disturbances move at almost the same speed as the former sea-breeze fronts.

Observations which illustrate some of these features were obtained in some radar observations of the collision in south-east England between a sea-breeze front and another similar meso-front (Rider & Simpson, 1968).

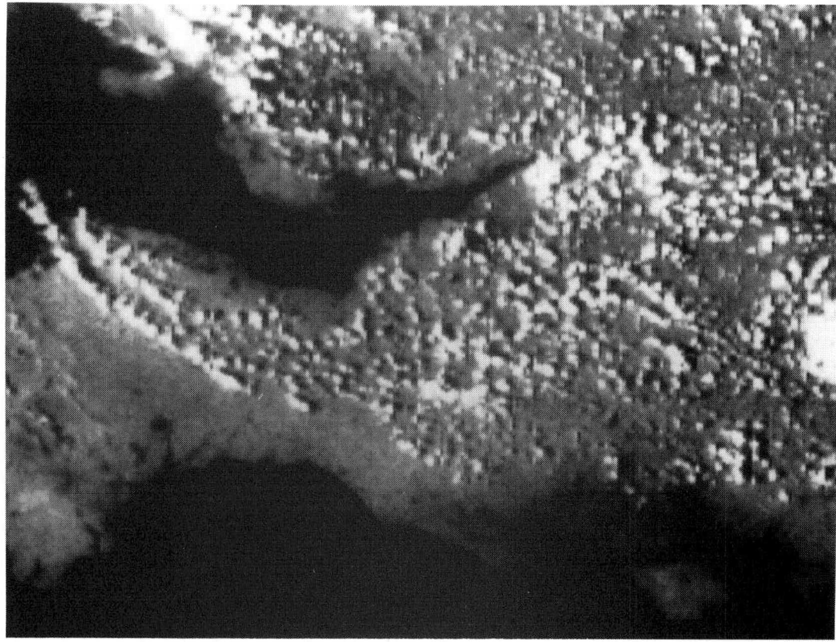

Figure 7.9. Double line of sea-breeze fronts north of Lyme Bay in south-west England on 8 July 1983. (Photo by University of Dundee.)

Fig. 7.10 shows the position of the two shear lines in south-east England. The more southerly line, 'A', was moving away from the cold high pressure area in the south, which was caused by a long period of rain in the slow-moving weather system. Line 'B' was a sea-breeze front, a weaker feature, moving from the coast shown near the top of the map.

The images of these two lines were followed by radar for two hours as they approached each other. They met at about 1415 GMT and then apparently 'crossed', since two echoes could be seen moving away from the line of collision. Fig. 7.11 shows four stages in the collision, in which the clear dark lines on the radar screen are thought to be due to swifts soaring and feeding at the shear lines, and the dotted areas are caused by rain.

After the collision the sharp upward bulge of air led to a line of instability moving south, which soon appeared on the radar as patches of rainfall, and also to a bore moving north. It seems that all the swifts transferred their allegiance to the line of the north-moving bore, which might be expected to produce a steady source of rising air carrying airborne insect food.

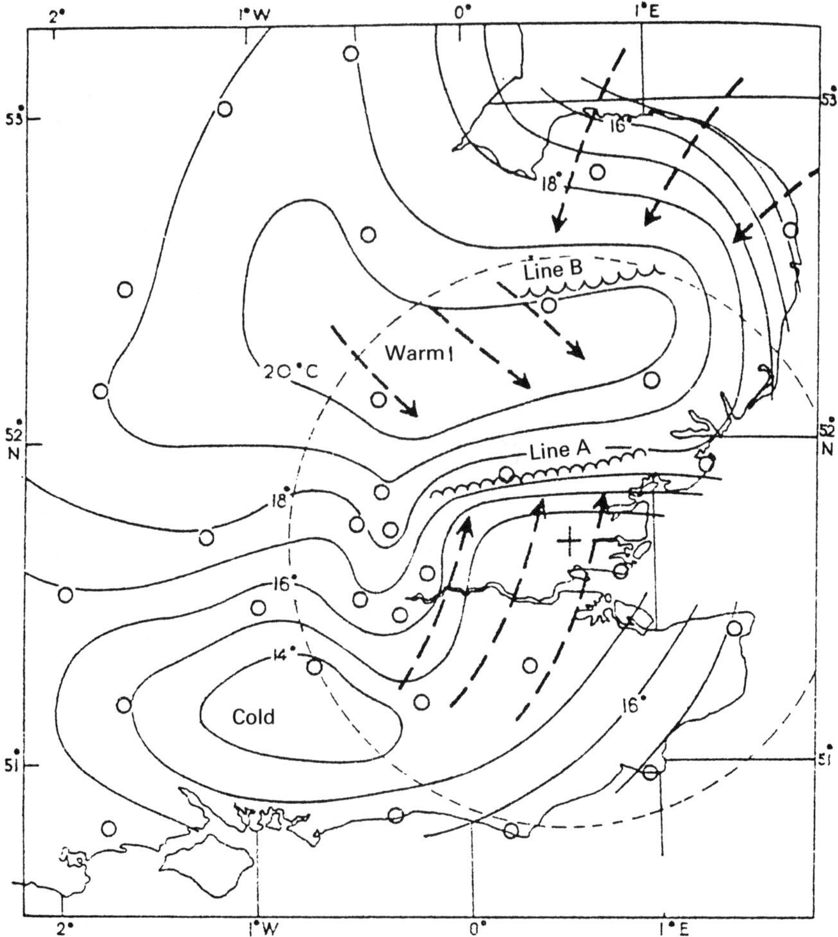

Figure 7.10. Two shear lines, labelled A and B, in south-east England, which were followed by 23 cm radar, as seen at 1200 GMT on 5 July 1966. Temperatures are marked at intervals of 1 °C and the arrows show wind flow. The dashed line is a circle of radius 50 miles from the radar station (shown by the cross). The open circles are observing stations.

7.52 Collisions observed by doppler lidar

Field measurements have shown that many thunderstorm outflows have similar characteristics to sea-breeze fronts (Moroz & Hewson, 1966). The frequent observations of these outflows are therefore of interest in the study of sea-breeze fronts.

Studies of colliding outflows have been made using both doppler lidar and a meteorological tower (Intieri, Bedard & Hardesty, 1990) in which three cases

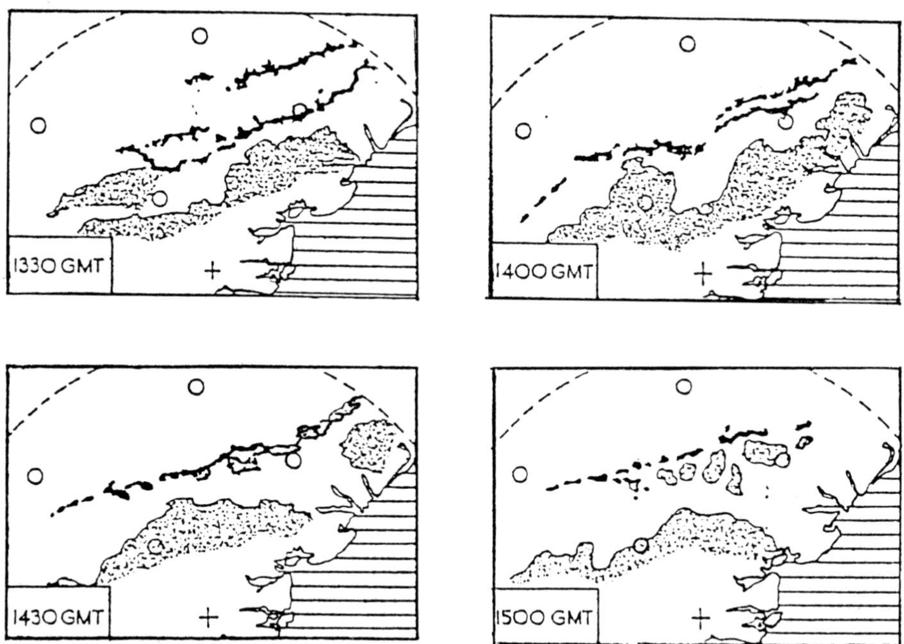

Figure 7.11. Four stages in the collision of two shear lines on 5 July 1966. The sharp dark lines seen on the radar are echoes from swifts, and the shaded areas are echoes due to rain.

showed the capacity of the lidar to observe the fine-scale details of collision interactions. Further details are given in Chapter 10. All these collisions resulted in new convection as the warmer of the two outflows was lifted over the cooler, denser one, confirming the results described in the previous section in which convection was seen by radar to result from the collision of a sea breeze and another front.

7.6 Islands

7.61 Malta

The island of Malta, which is 26 km long and 12 km broad, lies in the Mediterranean at a latitude of 36 °N. Between March 1952 and April 1954 investigations of Malta sea breezes were based at the Met. Office station on Luqa airport. As a result a good picture was built up of the formation of convergence zones over the island in different synoptic wind conditions (Lamb, 1955).

In the early stages of a strong sea-breeze day a cloud patch usually appears at a fixed position over the leeward side of the island. As the day progresses a convergence line develops against the general wind towards the centre of the island. The map in figure 7.12 shows the positions of convergence lines at the stage of maximum development for both north-westerly and south-easterly winds.

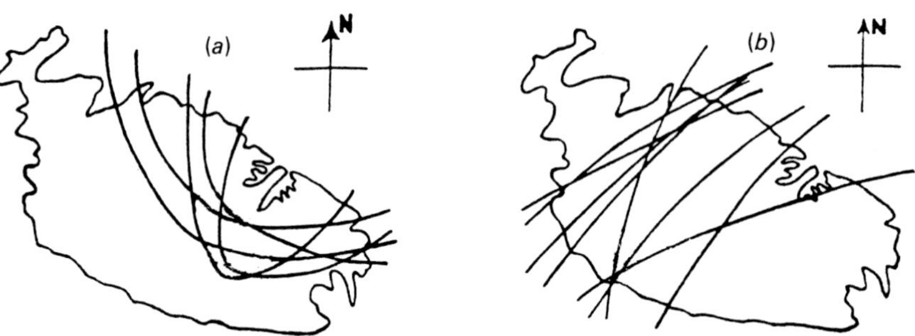

Figure 7.12. Positions of convergence lines observed on the island of Malta between 1952 and 1954. (*a*) Five cases with north-west gradient winds. (*b*) Eight cases with south-east gradient winds. (From Weather magazine, by permission of Royal Meteorological Society.)

On several days in August and September the cloud development was sufficient to produce a nearly stationary shower from large cumulus with tops reaching 3000 m.

In the later afternoon the convergence zone moves back towards the leeward end of the island and may pass out to sea. Here, evidence from sailing people suggests that the sea-breeze circulation is carried out to sea before it finally dies away.

The behaviour of the Malta sea breeze may serve as a simple illustration of tendencies likely to be at work in the convection pattern over larger islands, but the Malta case is essentially different from that of hillier or mountainous islands where coastal and anabatic breezes may be combined into one system.

7.62 Majorca

The island of Majorca in the western Mediterranean is larger than Malta: it is about 65 km square and has a mountain range near the northwest coast. The mountains have a mean altitude of 800 m, with a peak of 1448 m.

Nearly every day in the summer the synoptic winds are weak and the sea breeze is present. Using information obtained from farmers and fishermen the streamline map in figure 7.13 was compiled (Jansa & Jaume, 1946). This also shows the convergence zones which were observed. Some of these areas of convergence are due to the collision of sea-breezes from opposite coasts and the

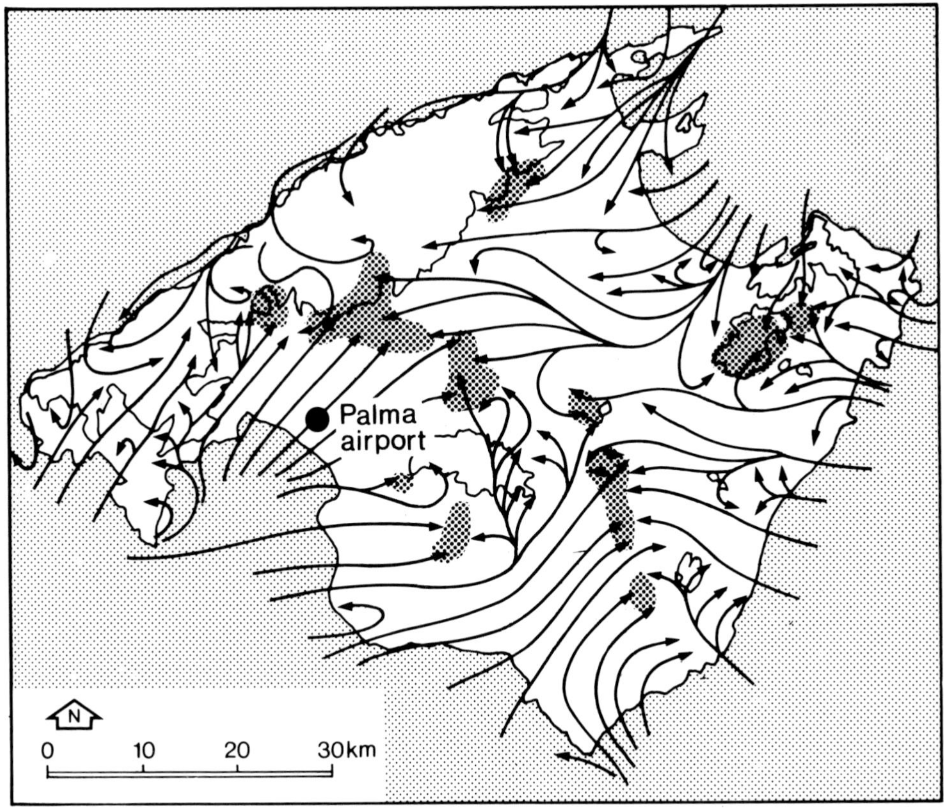

Figure 7.13. Streamlines of the sea-breeze observed on the island of Majorca. Shaded areas are convergence zones dues to collisions of sea-breeze fronts or to convergent flows at headlands. (After Jansa & Jaume, 1946.)

others are associated with the convergent flows at headlands. The development of these convergence lines has been confirmed in some satellite photographs taken by Meteosat on 11 May 1987 (Ramis & Alonso, 1988).

7.63 Larger islands

Investigations have been carried out into the influence of the land mass size on the intensity of sea-breeze development, using mathematical models. It was found that the sea breeze from a 'small' island, of radius 26 km, produced much smaller vertical velocities than a 'large' island of radius 50 km (Neumann & Mahrer, 1974). It has also been demonstrated that a circular island should produce stronger vertical velocities than an elongated island of equal width

(Mahrer & Segal, 1985). In a recent simulation of the effect of a strip of land on sea-breeze development; (Xian & Pielke, 1991) it appeared that the maximum sea-breeze convergence is obtained with a width of 150 km.

7.7 Land-breeze convergence

In contrast to the summer sea breeze, which leads to convergence over the land, the winter land-breeze circulation can lead to convergence over the ocean. Land-breeze convergence is responsible for the more intense type of Great Lake snow storm which forms at cloud-lines parallel to the coast and which may persist for some hours.

7.71 Land-breeze convergence on Lake Michigan

Using radar and aircraft, the winter land-breeze from one or both shores of Lake Michigan has been shown to have an important role in organising low-level convection (Passarelli & Braham, 1981). Figure 7.14 shows a cross-section of winds and potential temperatures recorded from aircraft flights. The Michigan land-breeze front is seen at 87.25 ° W; further west is the Milwaukee land-breeze front which was only detectable in the lowest aircraft pass.

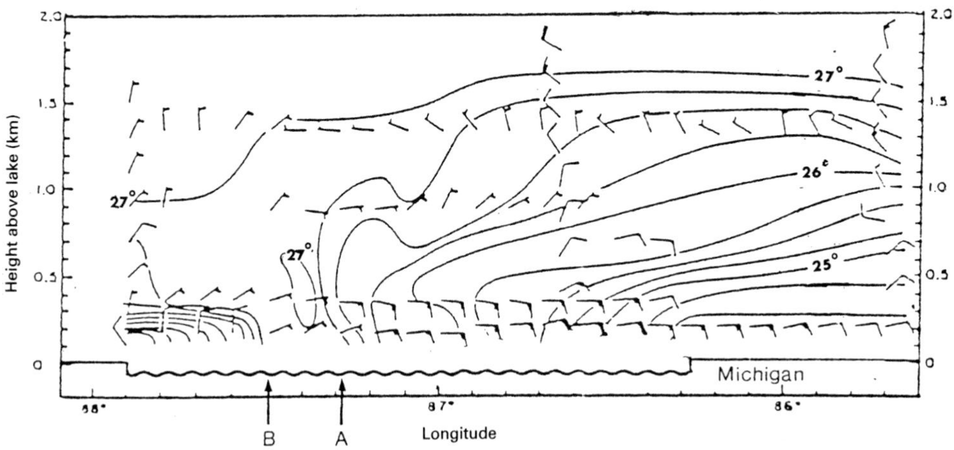

Figure 7.14. Land-breeze fronts over Lake Michigan on 7 November 1978 seen in east–west cross section. A marks the Michigan land-breeze front and B the much shallower land-breeze from the west coast. The contour lines show potential temperatures in °C, wind arrows are also shown. (From Passarelli & Braham, 1981.)

The continual heating of a land-breeze flow by the underlying warm lake generates a gravity current, which reinforces the offshore flow. The presence of the opposite flow is important since gravity currents from the two shores can oppose one another and generate strong low-level convergence as confirmed by the cloud pattern.

An outstanding example of a cloud band over Lake Michigan seen on a satellite picture (Elsner, Mecikalski & Tsonis, 1989) on 23 February 1989 appears in figure 7.15. This cloud band developed in the early morning and remained nearly stationary over the centre of the lake throughout the daylight hours. With the exception of the single cloud band, skies were clear over the lake.

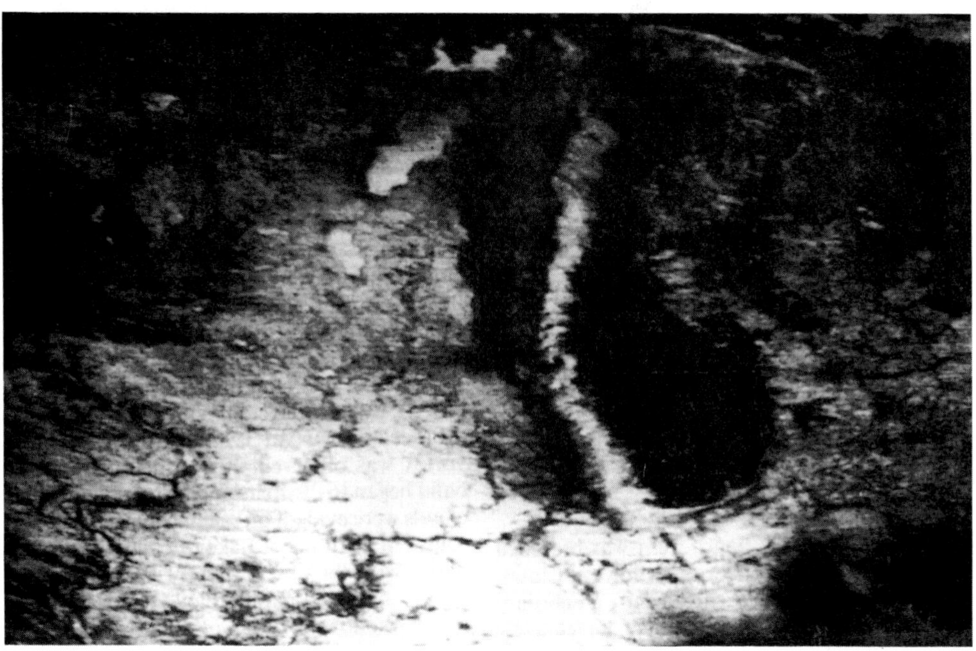

Figure 7.15. Satellite image of Lake Michigan at 1500 UTC on 23 February 1989. Clear skies are seen over the lake except for the cloud band, parallel to the shore, caused by land-breeze convergence.

(a)

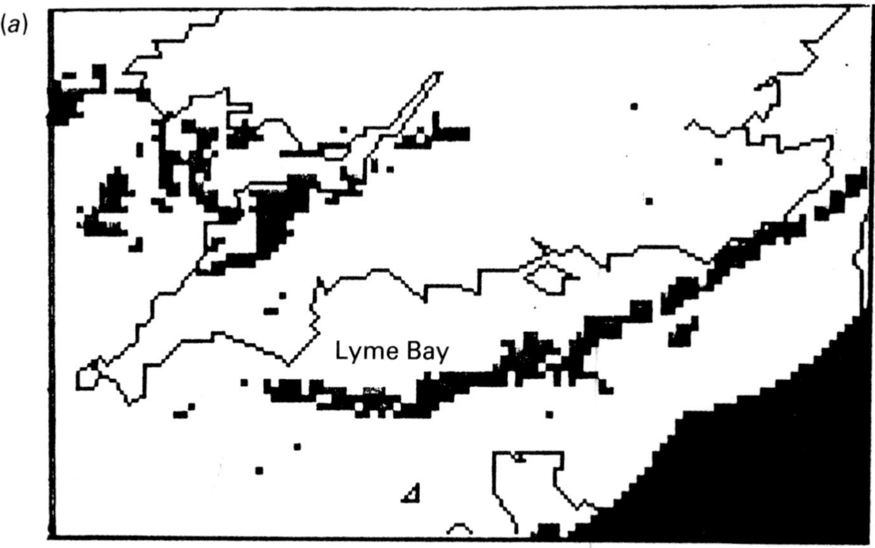

(b)

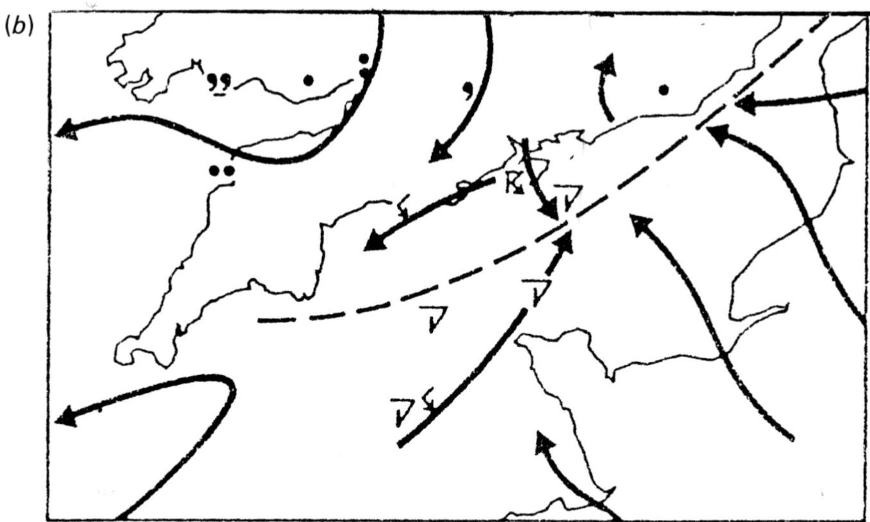

Figure 7.16. Land-breeze convergence zone at 0600 GMT on 19 October 1990, as shown by (*a*) rainfall intensity over the English Channel from UK weather radar network, and (*b*) The surface wind streamlines and present weather; the dashed line shows the position of the convergence zone. (From *Meteorological Magazine*, by permission of HMSO.)

7.72 Convergence over the English Channel

The image related to a convergence line over the English Channel shown in figure 7.16 is based on rainfall intensity from the UK weather radar network (Waters, 1990). It followed a night during which the first rain cells developed over Lyme Bay at 2300 GMT, thought to have been locally enhanced by the focussing of offshore land-breeze circulations.

8

Life and the sea breeze

The life-histories of airborne creatures are affected by many atmospheric systems and structures. The sea-breeze system, and especially sea-breeze fronts, have been shown to play an important part in the lives of birds and insects. Some of the latter which have become pests have been studied in great detail.

As well as insects, with their great economic importance, it has been shown that many birds make use of sea-breeze fronts.

The sea breeze also has significant effects on much smaller life forms, for example on the distribution of airborne pollen.

8.1 Pollen

Pollens and spores, as well as man-made particulates, are carried long distances through the atmosphere (Gregory 1961), and localised transport by the sea breeze is of interest for its effects at many coastal sites.

8.11 Sugar-beet and the sea breeze

The Agricultural and Food Research Council's site at Broom's Barn in East Anglia has monitored the behaviour of the sea breeze for many years. The significance here is the possible effect of the sea breeze in blowing pollen inland from the growths of wild sugar-beet which exist near the coast and its (undesirable) effect in pollenating cultivated beet crops further inland.

The sugar-beet plant, which forms a very important agricultural crop, is a biennial, producing a rosette of leaves and its large sugar-rich root in its first year. It does not normally flower until the second year, but a proportion of plants do become reproductive in the first year, a condition called 'bolting' in which the flowering stems grow very rapidly, by even as much as 15 cm in a day. This adversely affects the yield of the crop and produces seed; this heritable condition can be passed on through the pollen.

Infestations of 'weed-beet', whose origin is in seed produced by 'bolters', occur in sugar-beet crops. This problem arises from interpollinations from easy-bolting types in the cultivated plants, or from wild beet populations, and occurs in most sugar-beet growing countries, particularly in maritime areas. The map in figure 8.1 of the distribution of sugar-beet crops in England in 1969 shows that the majority of the crop lies in areas where sea-breeze penetration is likely. In the coastal regions of East Anglia where wild beet grows, pollen from wild beet may have fertilised crops, especially in the years 1971–1973 when the prevailing south-westerly winds were deficient at flowering time (Longden, 1987).

The most severe weed-beet infestations are to be found in Belgium, north-east France and Great Britain where much sugar-beet is grown in large areas on

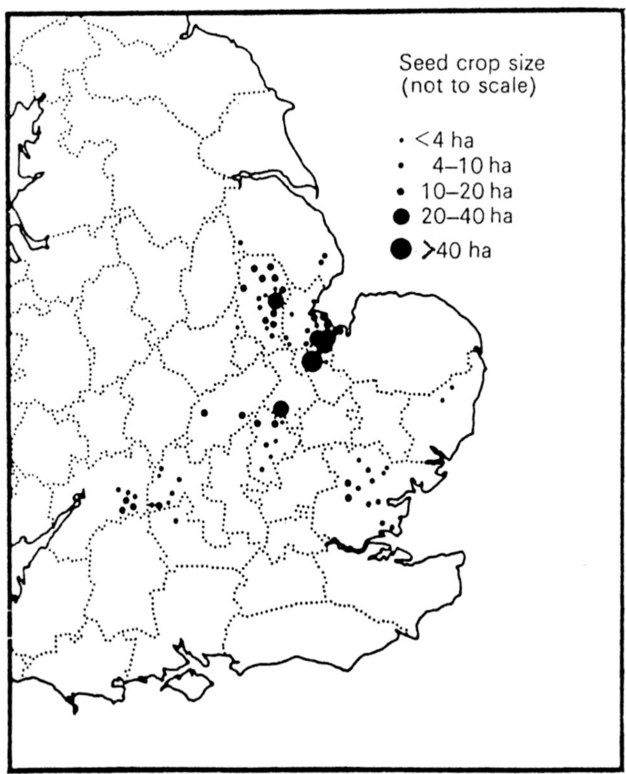

Figure 8.1. Location and size of sugar-beet seed crops of 1969 in England, showing their concentration near the coast. (From Longden, Scott & Tydlesley, 1975.)

close rotations with cereals in maritime areas; see figure 8.1, from Longden, Scott & Tydlesley (1975).

8.12 Airborne pollen at sea-breeze fronts

Measurements of airborne pollen over Long Island, USA, have shown some effects due to the sea breeze (Raynor, Hayes, & Ogden, 1974). Both vertical ascents and traverses through sea-breeze fronts were made with a sampler mounted on a light aircraft. Some pilot balloon data were also available. The area of these measurements is shown in figure 8.2.

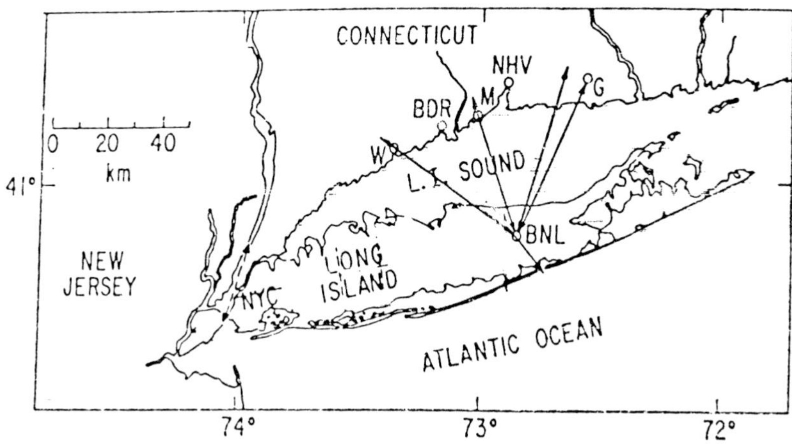

Figure 8.2. The Long Island area; the arrows are paths of sea-breeze flights in pollen collection. NYC, New York City; BNL, Brookhaven National Laboratory; BDR, NHV, W, M and G denote cities in Connecticut. (From Raynor *et al.*, 1974)

Flights were made in May and June when pines and oaks were pollinating, and later in August and September when ragweed pollinates. Ragweed pollen has a gravitational settling speed of 1.6 cm s^{-1}; most of the tree pollens are larger, with greater settling rates.

It was found that sea-breeze flows greatly modified profiles of pollen, by decreasing concentrations at low levels. Pollen was found to be carried aloft at the sea-breeze front and recirculated in the returning flow.

Figure 8.3 shows the data from one of the ascents through the sea breeze. During the ascent the top of the sea-breeze flow was near 1.5 km; the high pollen concentrations near the ground decreased to zero at 1 km height. By the time of the second ascent the sea breeze had greatly decreased in depth, with

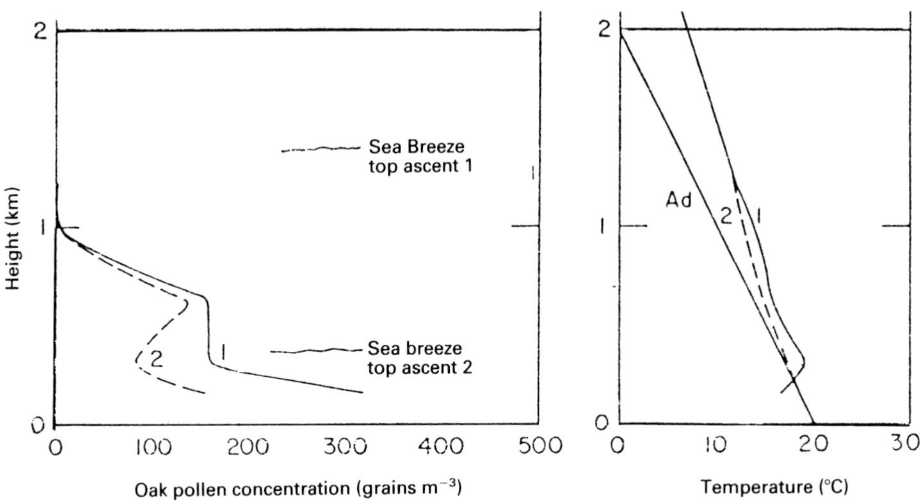

Figure 8.3. Pollen concentration and temperature profiles on ascents during sea-breeze flows. 1 and 2 mark the first and second flights on 18 May 1972, showing the secondary maximum in the second flight. Curve Ad is the adiabatic lapse rate. (From Raynor *et al.*, 1974.)

its top being at only 0.35 km. The pollen concentrations were low below this depth, but a maximum occurred at 0.6 km, presumably in the return flow.

The formation of sea-breeze fronts over Long Island can be complex, as shown in the measurements of figure 8.4 made during a traverse by an aircraft. A cold tongue of sea-air penetrates to the centre of the island and the pollen distribution shows high concentrations rising at the sea-breeze front. The secondary maximum is due to a local sea-breeze front from the sound in the north.

8.2 Insect pests

Many insects which attack food crops are seriously dependent on the behaviour of local winds. These winds transport them, concentrate their numbers and control their breeding and feeding behaviour.

8.21 Insects at converging local winds

The most familiar insect pest, going back to biblical times, is the desert locust. Locusts are large grasshoppers, which, although sometimes solitary, become gregarious from time to time, when their vast swarms are a serious plague.

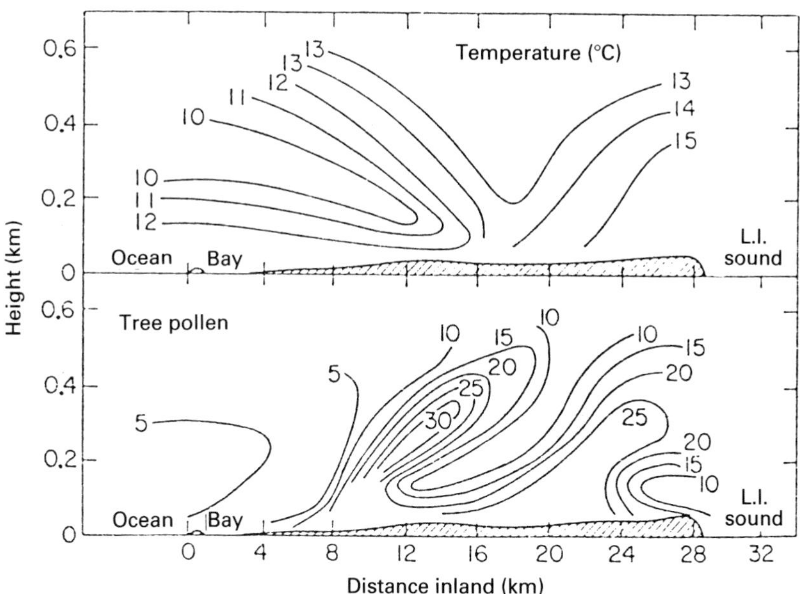

Figure 8.4. Cross sections of temperature and pollen concentration (in grains m^{-3}) over Long Island during a sea breeze on 6 June 1972. (From Raynor *et al.*, 1974.)

Locust swarms develop in Africa in the local winds of the Inter Tropical Convergence Zone (ITCZ) (Rainey, 1969). This forms at the convergence of the north-east trades and the southerly monsoon, across the entire continent of Africa, and moves north and south during the year.

The Inter-Tropical Front is often sharply defined and has been systematically explored by aircraft. An example is given in figure 8.5, which shows the position of Desert Locust swarms in a period when the ITCZ had remained effectively stationary. This map shows the position of Hargeisa in the Somali Republic where the photograph in figure 8.6 was taken. This shows the pattern formed by airborne locusts at the leading edge of a cold outflow, probably from a storm on the ITCZ.

A persistent zone of wind convergence significant in the distribution and breeding of the desert locust is the Red Sea Convergence Zone (Rainey, 1976), shown in figure 8.7, which is found almost continuously from October to May within the Red Sea basin. In this flight, made with wind-measuring aircraft, no insects were netted, but the presence of this convergence helps to explain some previous concentrations of scattered adult desert locust populations.

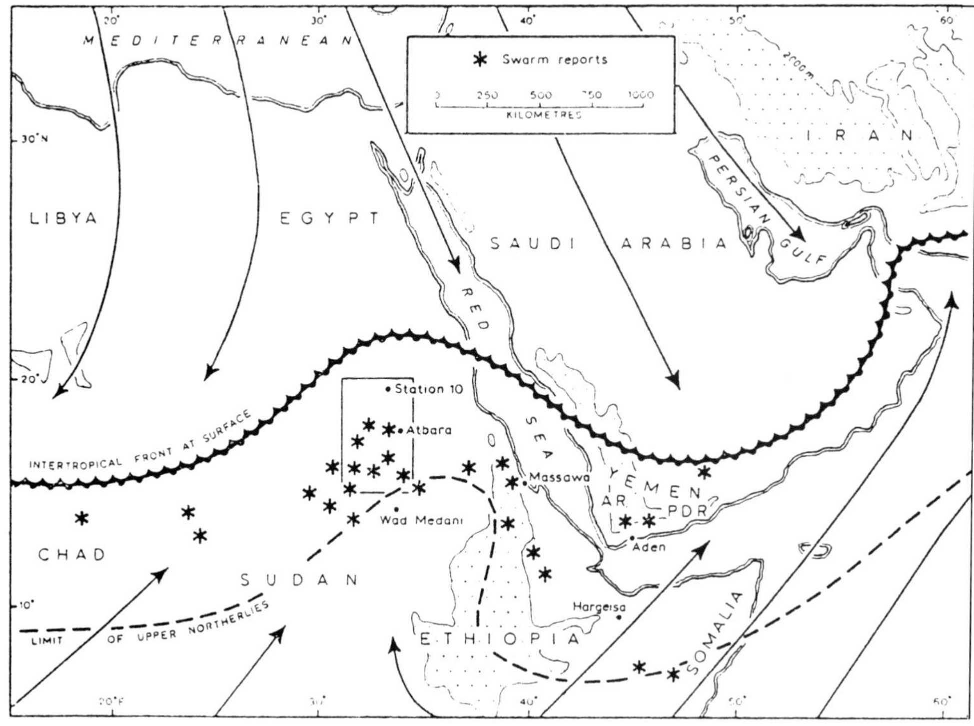

Figure 8.5. Desert locust swarms in the Inter-Tropical Convergence Zone during a period in July 1950 when the zone remained almost stationary. Arrows show general wind directions. (By permission of the Royal Meteorological Society.)

8.22 Land and sea breezes

The presence of a sea breeze offers advantages to airborne insects because it can prevent them from being flown out to sea and transports them back inland away from coastal areas.

Observations of insects coming in from the ocean at Sydney, Australia, where no breeding sites exist upwind, have been explained by the insects being carried out the previous night by the land breeze, or even in higher synoptic wind currents and then descending and returning to the coast in the low-level sea breeze.

8.23 Insects at sea-breeze fronts

The arrival of the sea breeze during the late afternoon at Canberra in Australia is well known to the city's inhabitants, where the cooler easterly flow brings

Figure 8.6. The form of a cold outflow outlined by flying locusts, seen at Hargeisa in the Somali Republic, 3 August 1960. (Courtesy of A.J. Wood.)

welcome relief. The Canberra sea-breeze echo on radar is caused almost entirely by flying insects, mostly grasshoppers, which the CSIRO radar is known to be very effective at detecting (Drake, 1982). The profiles of the leading edge of one of these sea-breeze fronts as it advanced are shown for a number of occasions in figure 8.8. These profiles, which are traced from the reflections from the oncoming wall of airborne insects, are similar to those already illustrated in measurements of sea-breeze fronts outlined by haze and from airborne measurements of temperature and humidity.

Further confirmation that this reflecting area represents the leading edge of the cold air comes from measurements of the distribution of airborne moths at a sea-breeze front in Canada (Schaefer, 1979). This sea-breeze front was at New Brunswich and the moths were those of the spruce budworm (*Choristoneura fumiferana*). Although the numbers of moths involved were small, airborne radars collected much information about the sea-breeze frontal structure.

A sharp windshift occurred at a point determined by the Doppler navigation system carried by a DC3 aircraft which was equipped to measure airborne insect concentration. Figure 8.9 shows temperature, turbulence, wind and insect density measured on a traverse of the front, from which the sharp boundary of the high-density airborne moths is apparent.

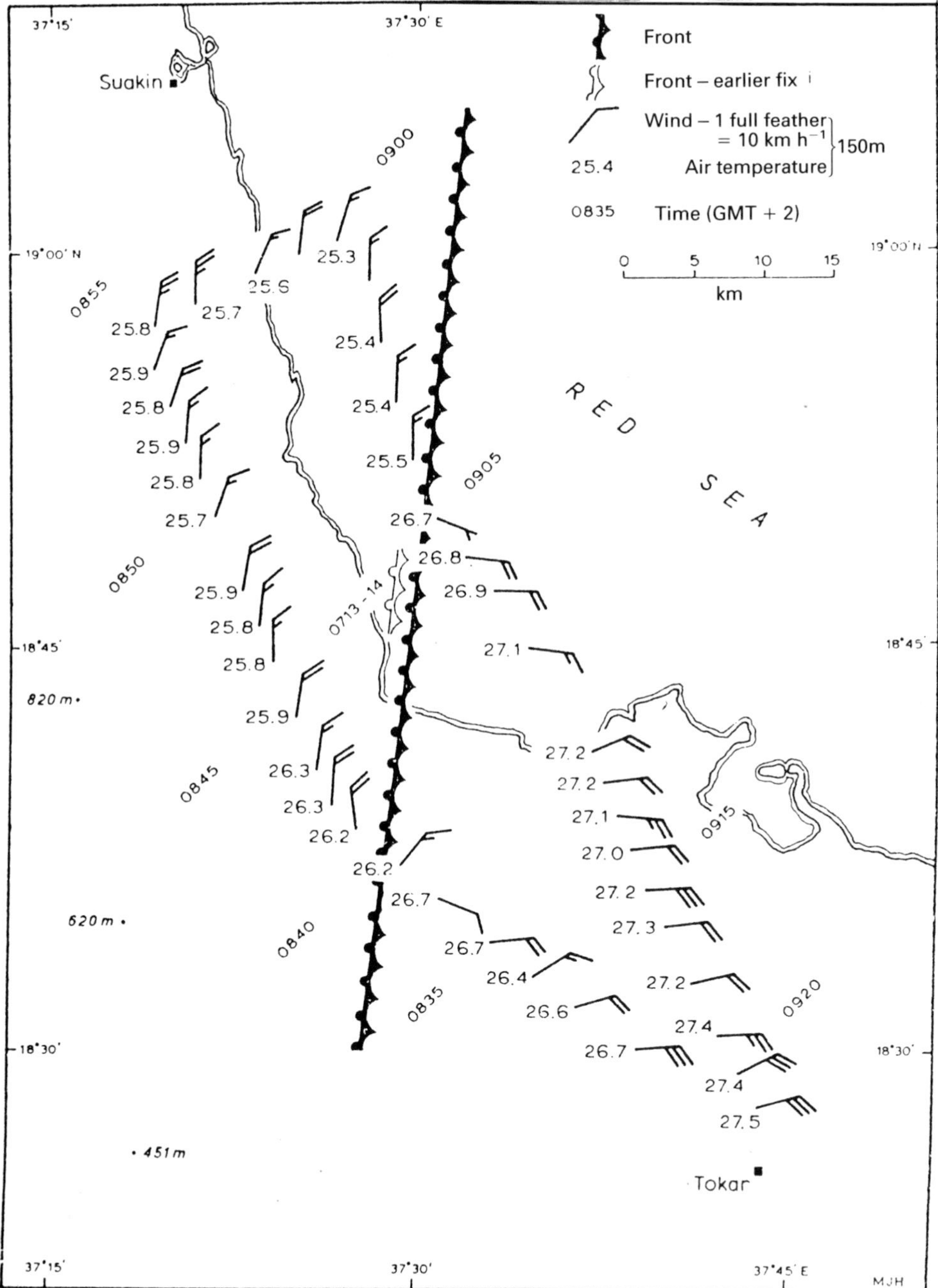

Figure 8.7. The Red Sea Convergence Zone on 17 November 1970. The wind convergence concentrated airborne insects. (By permission of the Royal Meteorological Society.)

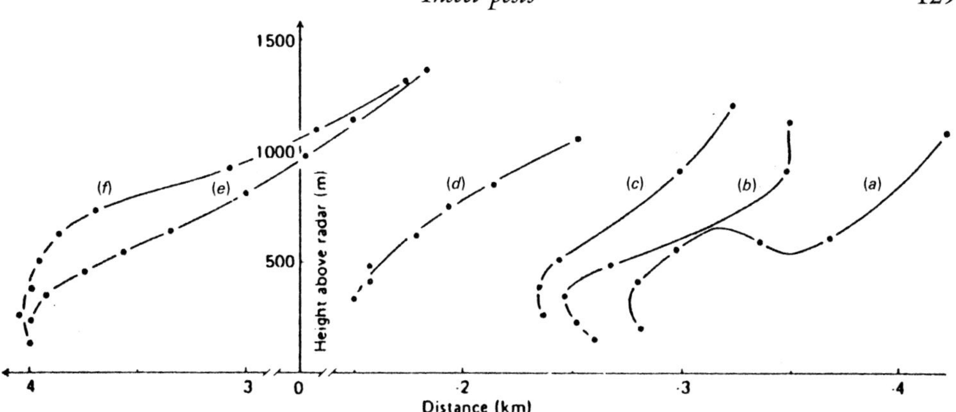

Figure 8.8. Profiles of the leading edge of the sea-breeze front at Canberra, Australia. Radar echoes are from flying insects, mostly grasshoppers. (*a*), (*b*) and (*c*) in 1921; (*d*) 1930; (*e*), (*f*) 1942. (From *Weather*, by permission of the Royal Meteorological Society.)

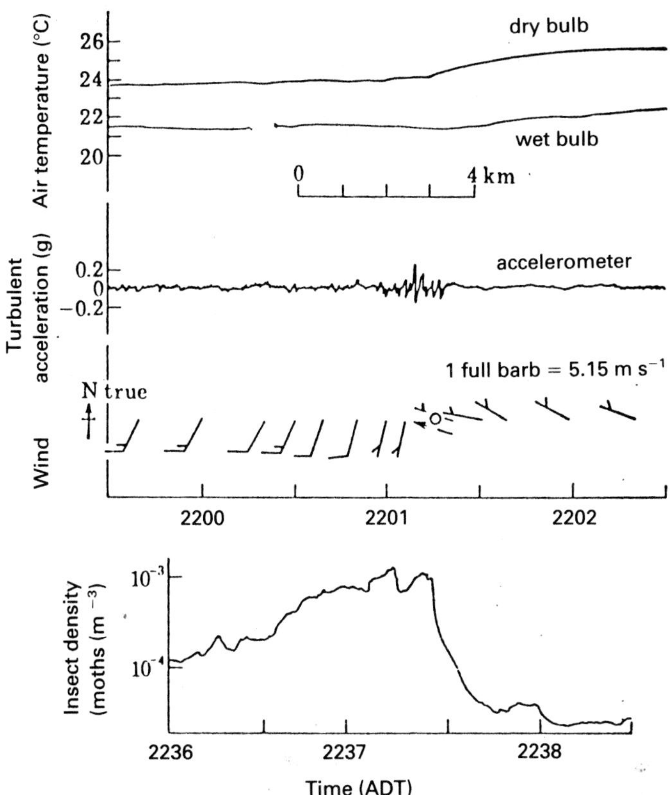

Figure 8.9. Temperature, turbulence and wind records from a sea-breeze front at New Brunswick on 10 July 1976. The insect density (moths) near the front is also shown. (After Schaefer, 1979.)

8.24 Concentration of airborne insects

Mechanisms which concentrate the numbers of migrant airborne insects have been thoroughly studied (Pedgley, 1990). We have already described the detection of dramatic increases of insect numbers at convergence zones; in such a zone, as shown in figure 8.10, the horizontally converging winds must be accompanied by a vertical flow of air. If the concentration of insects from the left is 'a' per cubic metre and from the right is 'b', then after rising and mixing the concentration is $(a+b)/2$. This is on the average no more than it was before.

However, there are more insects above any given area on the ground than there were earlier and if any fall-out exists then the volume concentration can increase. If fall-out is significant then the insects are no longer 'neutral tracers' of airflow.

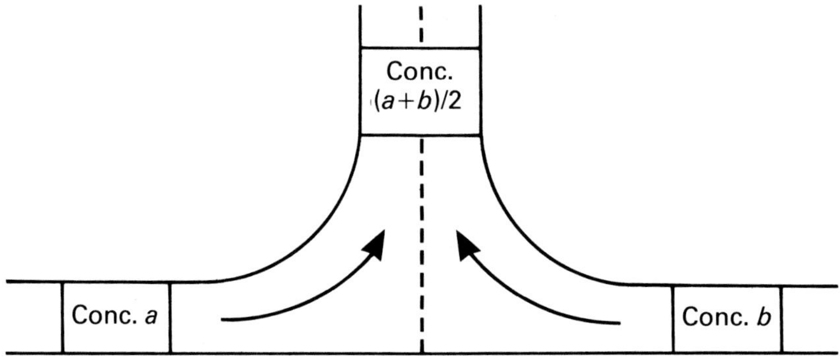

Figure 8.10. Converging winds may not increase insect density if no fall-out exists; the new density is only the mean of the two original densities.

There are good reasons why fall-out can be expected. The insects may descend because of the temperature decrease they experience when carried upward. The rate of cooling of 10 °C per 1000 m can affect flight characteristics; they may be less efficient at maintaining height, they may even make deliberate efforts to descend. There may also be visual stimuli which will reduce height, or descent may be programmed to occur after a certain time of flight.

So we conclude that aerial concentration by convergent winds is possible if at least some of the insects resist being carried aloft, either by reducing their lift or by just falling.

8.25 Numerical model of concentration of airborne aphids

Aphids are the outstanding airborne insect pests in Britain in terms of their economic importance. The bean aphid, *Aphis fabae*, can sometimes cause catastrophic damage to spring-sown field beans. After winter spent on their host plant, the spindlewood tree, in May and June the aphids migrate to beans and other crops. Both in southern England and in East Anglia unexpectedly large infestations have occurred, considering the numbers known on their host plants. It seems likely that one of the reasons for this is the massive transport of winged aphids by the inland progress of sea-breeze fronts. In view of this, the process of insect concentration at a sea-breeze front has been modelled numerically by Mansfield, Milford & Simpson (unpublished).

The flow field used in the model was from actual measurements of an almost ideal sea-breeze front, likely to be typical of the penetrative sea breezes. With simple assumptions on insect behaviour it can be shown that such a flow field may have a significant effect on their distribution.

The behaviour imposed on the insects is that they are originally flying in the bottom 100 metres, they have negligible horizontal speeds and their vertical rate of sink increases linearly with height. This is based on the observations of Berry & Taylor (1968) that aphids may climb at 0.2 m s^{-1}, or fly down at 0.6 m s^{-1}. If incapacitated by low temperature, when inert they can fall as fast as 0.9 m s^{-1} with wings out or twice this speed with wings folded.

In the computational procedure, each 'box' containing insects measures 500 m horizontally and 100 m vertically. At intervals of 50 seconds each box is moved along the relevant streamline and the insects contained are redistributed.

The results after 20 000 seconds are shown in figure 8.11. Figure 8.11(*a*) was calculated with no fall speed and can be compared with figure 8.11(*b*) which included the graduated fall speed. It is clear that the insects are present near the front in vastly greater numbers in the case with fall-out than in the case with no fall-out.

The final illustration in this section, figure 8.12, shows the concentration of spruce budworm moths measured by downward directed radar at the sea-breeze front in New Brunswick on 10 July 1976. The pattern is similar to that shown by the locusts photographed at Hargeisa (figure 8.6), with the highest concentration sweeping up along the leading edge. Without more information, however, it is not possible to say which of the two model results these data most resemble.

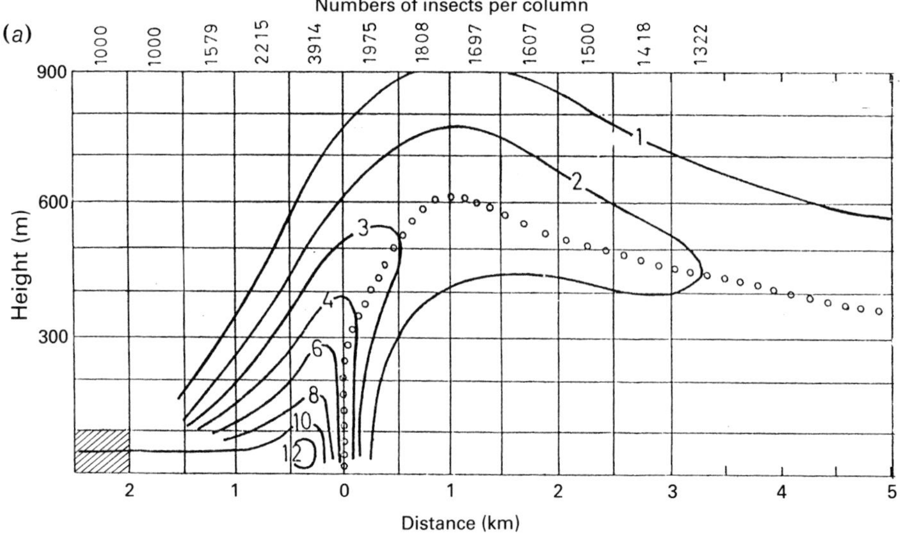

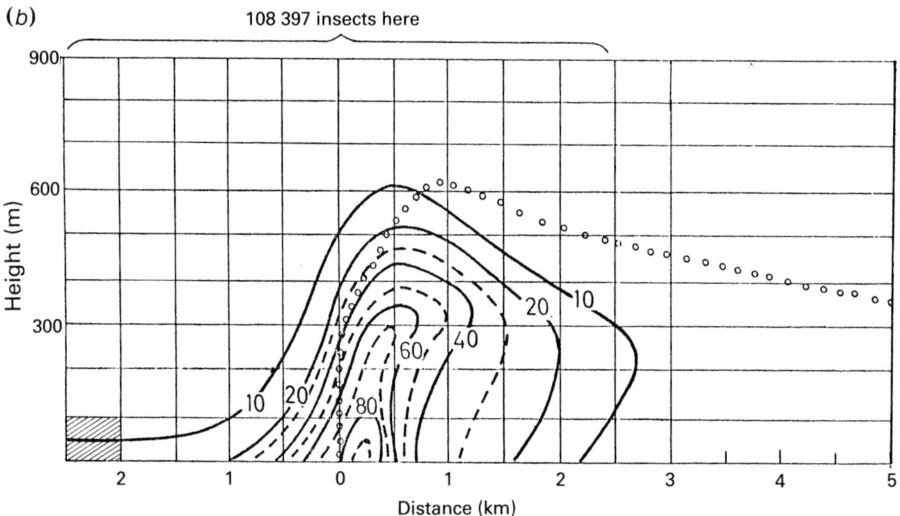

Figure 8.11. A numerical model of aphid concentration at a sea-breeze front. A continuous input exits near the ground and the results are given after 20000 seconds. The 'box' containing insects at the beginning of the model is shaded. Numbers of insects (in hundreds) are given on the contour lines. The line of circles denotes the sea-breeze front. (*a*) No fall-out; no increase in concentration. (*b*) Graduated fall-out; a greatly increased concentration.

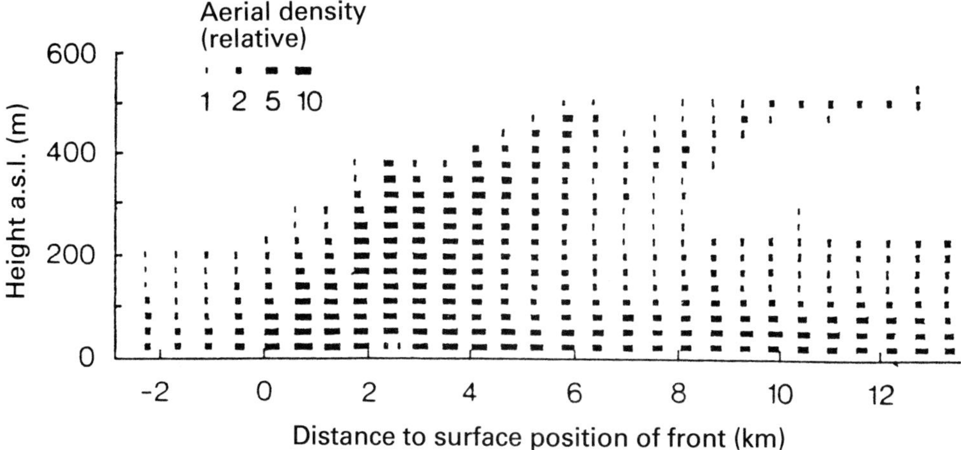

Figure 8.12. Density of flying moths at a sea-breeze front, seen from airborne radar traverse. A high concentration can be seen, sweeping up along the leading edge of the front. New Brunswick, Canada, 10 July 1976. (Courtesy of K. Allsop.)

8.3 Birds and the sea breeze

Many sea-birds soar on cliffs, using the wind from the sea, at their nesting sites. The sea breeze proper, brought about by the diurnal changes in land–sea temperature differences, plays an essential part in the feeding habits of a few specialists. The chief of these in Britain is the swift (*Apus apus*).

8.31 Swifts

The swift is an almost wholly aerial bird, it feeds on airborne insects, mates in the air and even collects its nesting materials in flight. It is black, with white underneath the head, and has long tapered wings, of span 45 cm.

The flight of swifts is distinctive, of an irregular darting pattern. Much of the time the swift does not seem to flap its wings at all, but just twitches them as it darts from left to right in feeding flight. In fact, at one time many people believed that swifts flapped their wings alternately. Careful records have shown that this is not so (Lack, 1956).

Swallows (*Hirundo rustica*) are similar in some ways to swifts, but they have a long forked tail and a red patch under the chin. These two birds have often been confused, but with practice they can be distinguished in the air by their different modes of flight – the swift makes rapid wing beats with short glides, the swallow beats with intermittent wing folding.

The two birds have developed a difference in their general feeding strategy; swallows fly mostly close to the ground when feeding, but swifts are often at heights where they cannot be seen from the ground.

Two other similar species are the house martin (*Delichon urbica*) and sand martin (*Riparia riparia*); these four birds are illustrated in figure 8.13.

Figure 8.13. A swift (*Apus apus*) (top), the swallow (*Hirundo rustica*), house martin (*Delichon urbica*) and sand martin (*Riparia riparia*). (Courtesy of Robert Gillmor.)

8.32 Swifts and sea-breeze fronts

Swifts feed entirely on small insects and spiders, which are caught in flight. In fine weather they often feed from 10 to 50 m above the ground but on still summer days they often circle much higher, occasionally to 500 or even 1000 m, anywhere where insects are numerous. When feeding the young, each meal brought back to the nest weighs just over a gram and consists of 300 to 1000 insects (Lack 1956). They may take up to an hour to collect. The record, which was measured at Oxford, consisted of 42 flights and totalled some 20 000 insects.

Swifts have learned to collect their airborne food where it has risen from the ground and is most concentrated. They use 'thermals' where they have been seen by glider pilots. Their habit of feeding in the neighbourhood of thunderstorms has long been observed from the ground, and for this reason swifts are sometimes known as 'thunder-swallows'. Glider pilots have spotted them circling in thermals and have also joined them in the rising air at sea-breeze fronts, where swifts often cruise along, feeding together in large groups.

Radar has been used to trace sea-breeze fronts; an early example from 1966 is shown in figure 8.14. This sea-breeze front was believed to be outlined by

Figure 8.14. Sea-breeze fronts in south-east England outlined by radar on 2 June 1966. The large dots are echoes from aircraft and some of the smaller ones are from groups of swifts. The arrow points to my aircraft, flying along a line of swifts at the front. The radius of the circle is 150 km. (Courtesy of Marconi Limited.)

the swifts feeding along it (Eastwood & Rider, 1961). At that time it was not clear whether high refractive index gradients in the atmosphere or the presence of birds was the cause of the echo. In the following years measurements were made of swifts in the air at other sea-breeze fronts by glider pilots soaring with them; there were usually enough swifts to give the required echo (Simpson, 1967). I used to meet them flying at heights between 600 and 800 m, often in groups of three or four, but on one occasion I passed over 200 in less than a minute.

8.4 The sea breeze and humans

The sea breeze throughout the world is known by local fishermen, who use the onshore breeze in the morning to return with their catch after a night's fishing. Other aspects of the sea breeze are also important to men and women.

8.41 The sea breeze and health

The sea breeze has usually ranked high among health-giving attributes of the sea-side (Bilham, 1934). However, some people in Jane Austen's time had reservations:

> My dear little creature, do not stay at Portsmouth to lose your pretty looks. Those vile sea-breezes are the ruin of beauty and health. My poor aunt always felt affected, if within ten miles of the sea, which the Admiral of course never believed, but I know it was so.
>
> (Austen, 1814.)

Nowaday a pale complexion is no longer so highly prized and the increased amount of sunshine on coastal resorts is generally held to be an advantage.

Because a sea breeze is southerly on a south coast and easterly on the east coast and so on, in general these winds have different characteristics. For example, air on the east coast of England is found to be 'bracing' and that on the south coast is considered to be 'relaxing'.

The clean sea breeze is always welcome; in some parts of Australia it is known as 'The Doctor' when the cool wind arrives on a hot evening miles away from the coast. The black side of the sea breeze, when photochemical pollution is produced in the sea-breeze circulation, has been dealt with in detail in Chapter 6.

8.42 Wind catchers and the sea-breeze

Some parts of the world are only habitable in houses with forced ventilation, which in some places in the Near East is obtained by using scoops in the roofs to carry the wind down into the buildings. This technique has a very ancient history as it is known that scoops were built in the roofs in the New Kingdom in Egypt facing north to collect the sea breeze. Figure 8.15 is a copy of a papyrus from about 1500 BC, showing houses on the roofs of which wind scoops can be clearly seen (Roaf, 1992).

Marco Polo, from a port in the Persian Gulf, comments

> 'The climate is excessively hot – so hot that the houses are fitted with ventilators to catch the wind. The ventilators are set to face the quarter from which the wind blows and let it blow into the house'
>
> (Latham, 1958).

Figure 8.15. A house with two wind catchers drawn on papyrus, found in the tombs of the New Kingdom in Egypt around 1500 BC. (Courtesy of S. Roaf.)

As the trade routes became established during the seventeenth century travellers described wind catchers at the gulf ports, each of which evolved its own particular design. One example is at Banda Abbas, where large square scoops are still built facing the sea breeze, with a palm frond mat over its vent which is soaked with water on summer afternoons.

At some places inland, where the prevailing seasonal wind changes direction, different types of wind scoops or wind towers are built. These multi-directional wind towers have air vents at the head and are square in plan, divided internally by two cross partitions so that air blows down between two upright planes on one side, and rises in the opposite space. Such wind towers, called '*badgir*'s, (literally 'wind catchers') are a characteristic feature of the architecture in the city of Yazd in Iran, where they are often sculpted and decorated. (Roaf, 1992)

The performance of such a wind catcher tower in Yazd is shown in figure 8.16. The air flow is sketched in and measured values of temperature and humidity are plotted for the later afternoon. Life in these conditions is tolerable when carried out entirely in the basement.

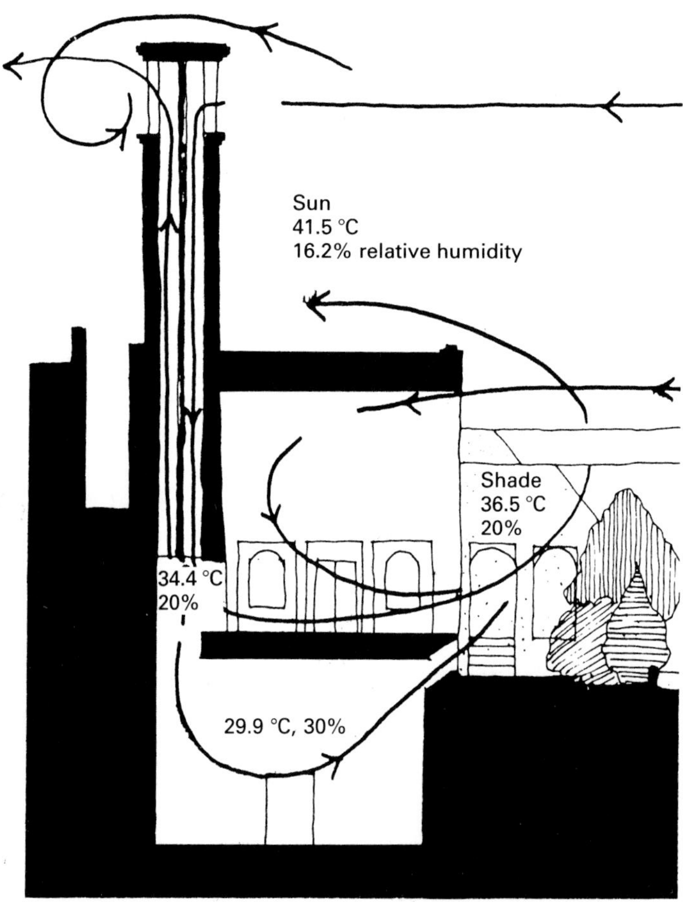

Figure 8.16. The performance of a wind catcher in a Yazdi house; life in these conditions is tolerable when carried out in the basement. (Courtesy of S. Roaf.)

In the city of Herat, further east in central Afganistan, the wind catchers are again uni-directional, facing north, and almost every room contains one. The rural dwellings are built up of cellular one-room units, which are cubes with sides of about 3 metres in length, each of which has a simple dome containing a wind scoop. Figure 8.17 shows how these units are built to form a typical two-story building (Hallett & Samizay, 1980).

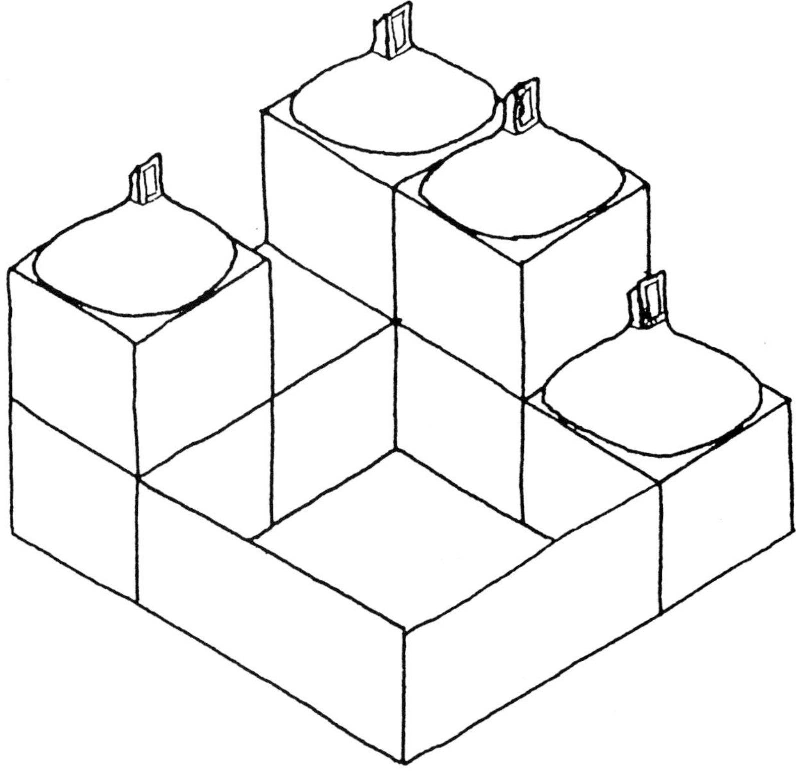

Figure 8.17. Rural dwelling near Herat, Afghanistan. Each section has a dome containing a wind catcher facing the prevailing local wind. (After Hallett & Samizay.)

In Sind, in West Pakistan, the sea breeze is well known locally for its tempering effect on the fierce summer heat of this arid region. The breeze blows from March to October with such constancy that the skyline in Hyderabad, 170 km inland, has an array of wind catchers built above the houses to receive the cool wind when it arrives. The sea-breeze front often arrives between 1700 and 2000 h with sudden severe squalls, and flying dust. There is usually no cloud, but a temperature reduction of 5 °C maybe expected. (Holmes, 1972) As figure 8.18 shows, the wind catchers of Hyderabad are designed quite differently from

Figure 8.18. Houses with wind catchers facing the sea breeze at Hyderabad, Sind. (From Hürliman, 1927, p. 261.)

those seen Afganistan and Iran, being wooden towers, with a plate fixed above at an angle of 45 °. In spite of some individual differences in design, it is clear that many people in dry arid regions rely on the use of the sea breeze and other local winds to make conditions bearable in the hottest weather.

9

Sports

Several outdoor sports can be affected by the sea breeze and for those which rely on the wind the arrival of the sea breeze has to be taken seriously into account. Such sports include gliding, ballooning and sailing.

9.1 Gliding and the sea breeze

On a fine summer day the extent of the land covered by the cool moist air of the sea breeze may become quite large. As the influence of the sea breeze extends so does the layer of stable air in which the thermals are either much reduced or even completely absent. Thus the area covered by the thermals so vital for the glider pilot steadily decreases. In a country the shape and size of England this can mean a serious reduction of the districts where soaring is still possible.

A good example of the movement inland of the sea breeze from all coasts is shown in figure 9.1. On 9 June 1968, a day of light winds and very good conditions for gliding, many long flights were made and a number of records were broken. The map shows the position of the sea-breeze fronts at 1800 GMT, when they had moved inland nearly 50 km. A number of memorable gliding flights are also depicted; they include a number of 300 km triangles and the first flight of 500 km to be completed in Britain. The areas selected by Brennig James for his flight remained unaffected by the sea breeze, but Ray Foot, flying further south, was not so fortunate. The sea breeze had swamped his course with stable air and no thermals were available to complete his return to Compton Down.

On the other hand, the rising air at sea-breeze fronts can be useful for soaring; many flights, either by design or by accident, have made use of this source of lift. The behaviour of the sea breeze is well understood by experienced pilots, who can often detect the position of the sea-breeze front from the marked

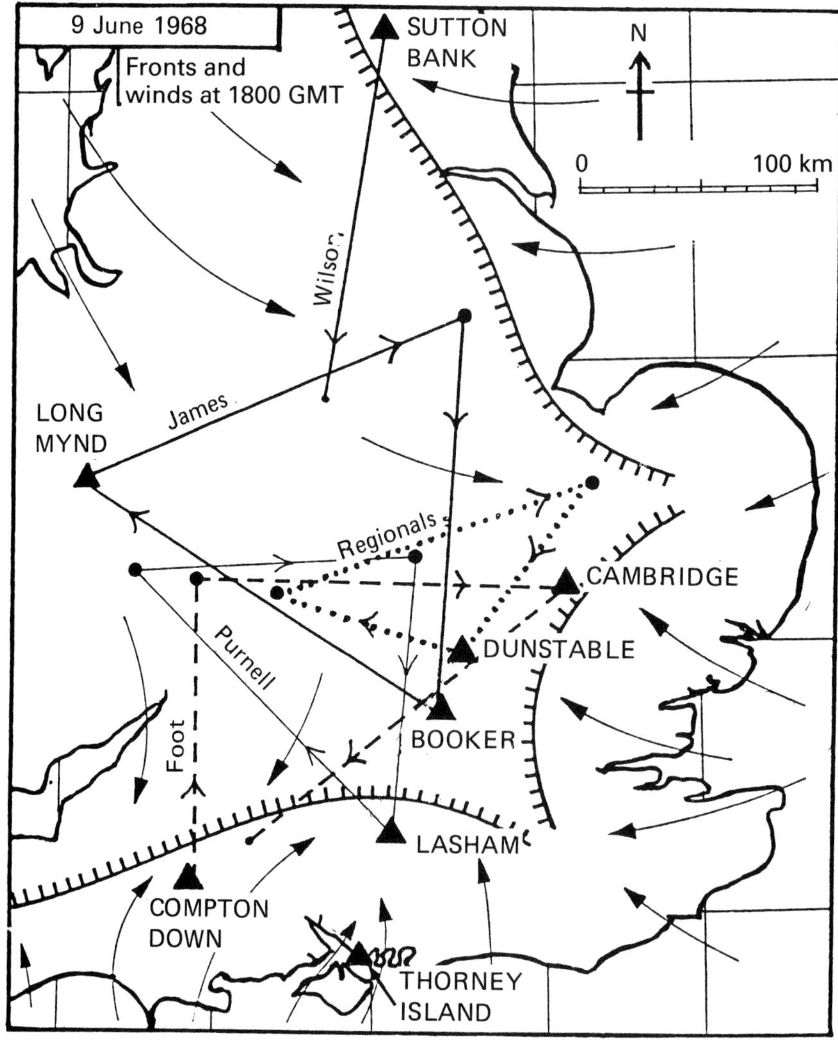

Figure 9.1. The spread of the sea breeze in England on a day of light winds. The lines show tracks of glider flights in the area in which soaring remained possible.

difference to be seen in the cloud forms. The main features seen are the 'curtain clouds', with a lower base than normal, and sometimes the greater depth of convection along a line of cloud. Even on a cloudless evening a weak line of lift at such a front has been used to complete the final 30 km to Lasham airfield, 50 km from the south coast of England. Figure 9.2 shows curtain clouds at a sea-breeze front seen from a glider.

Figure 9.2. Curtain clouds at a sea-breeze front, seen from a glider at 700 m on the landward side. The higher cloud-base of the cumulus clouds can be seen above the land. (Courtesy of Lorna Minton.)

9.11 Gliding and other convergence zones

Convergence zones described in Chapter 7 included some in California formed by the meeting of two sea breezes which had flowed round the mountains. One of these, the Elsinore shearline, has been regularly used by members of the local gliding club; a sketch of the formation is given in figure 9.3.

A 'sea breeze' over Central England (Diver, 1973) was observed by all the competitors in the National Gliding Championships at Husband's Bosworth, near Kettering in southern England on 13 June 1971. A pseudo-sea-breeze front developed in the clear air on the south side of a large stratus and fog area, with the cloud edge acting like the coastline in the normal sea breeze.

The sharp front lay across the desired route from Husband's Bosworth to Oundle Church, the first turning point of the day's task. There were good thermals at the start and pilots could see the edge of the frontal cloud, with typical curtain clouds ahead of it, as shown in figure 9.4. It was possible to climb at 3 m s^{-1} in the cloud at the front, but a 'dog-leg' had to be flown in order to photograph the turning point at Oundle Church.

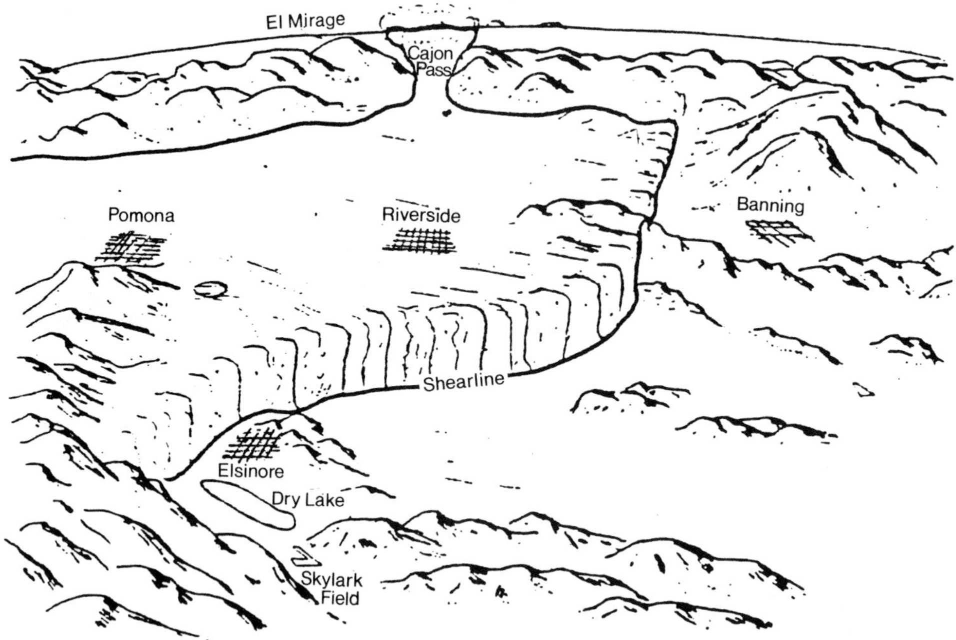

Figure 9.3. The Elsinore shearline, made visible by smoke and haze. Typical afternoon position as seen by glider pilots, looking north from above the Santa Ana Mountains. (After Lambie, 1963.)

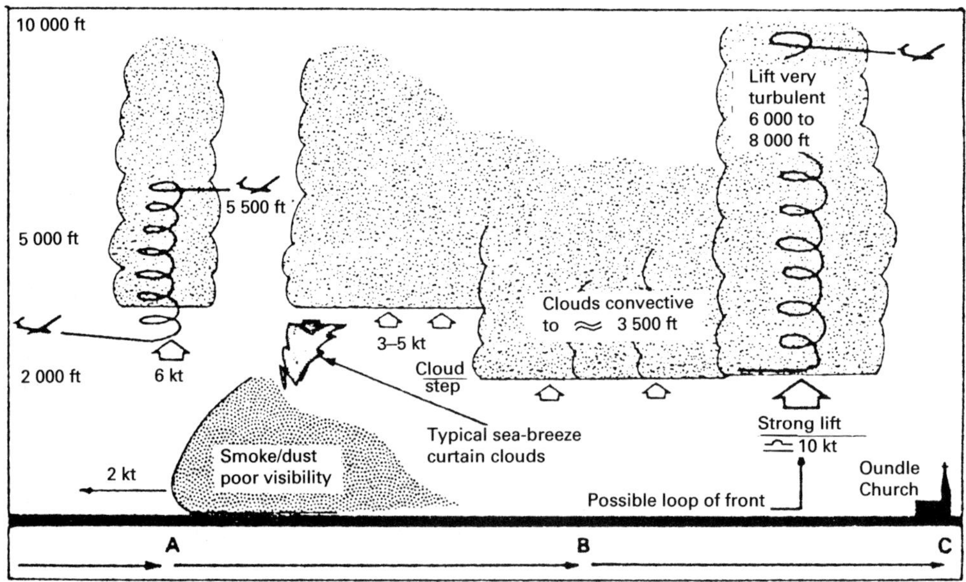

Figure 9.4. A pseudo-sea-breeze front (or 'cloud-breeze front') in the clear air at the south side of an area of stratus and fog. Seen by glider pilots from the National Gliding Championships near Kettering in central England on 13 June 1971.

Other isolated examples of convergence zones have been met by glider pilots and they would do well to bear in mind their possibility during flight planning, especially when large areas of overcast sky have a sharp boundary.

9.2 Ballooning and the sea breeze

Ballooning is now primarily a sport but it has a proud history of flights of adventure and scientific research. A hundred years ago a few expert balloonists were getting themselves involved personally in the sea-breeze circulation – sometimes it must be said, involuntarily.

9.21 Flights during the nineteenth century

An early deliberate use of the sea breeze was made, in 1861, during the American Civil War by the balloonist John La Mountain when he made reconnaissance flights for the Union Army (Rolt, 1966). Most of his flights were made in captive balloons, but on several occasions at Fortress Munroe, on the New Jersey Coast, he made free balloon flights over the enemy lines and returned to base. He did this by starting low down in the sea breeze, often at a height of less than a hundred metres. After having gathered his information, he was able to return to the coast at a greater height in the over-riding westerly wind.

In the late nineteenth century, still in the early days of ballooning, every flight could be an adventure in scientific research. A good example of this is a flight by Tissandier on 12th August 1868 in the sea-breeze circulation near Calais. He describes this in detail in the fascinating book *Travels in the Air* (Tissandier, 1871). At first, the crew were carried eight kilometres out to sea at 1700 m by a north-east wind; fortunately they had noticed clouds beneath them travelling in the opposite direction, and by descending to 400 m they were able to regain the shore and to land a few miles inland from the coast. After fixing their position, near the Calais–Boulogne road, an over-enthusiastic release of ballast took them right up into the clouds, and they were again swept out to sea. This time they were in serious trouble because the sea breeze had made its usual evening veer in direction. As a result they only just missed being swept away over the Atlantic, when they managed to land in a great hurry on the rocks near Gris Nez. After this flight, Tissandier reflected on his good fortune in recognising the opposite directions of the two currents of air, as shown in his illustration reproduced in figure 9.5, and concluded that 'this points clearly to what might be done by balloon were we possessed of a thorough knowledge of the direction of the winds'.

Two young adventurous balloonists, Eloy and Lhoste, resolved to celebrate the centenary of ballooning by crossing the English Channel. In a series of flights

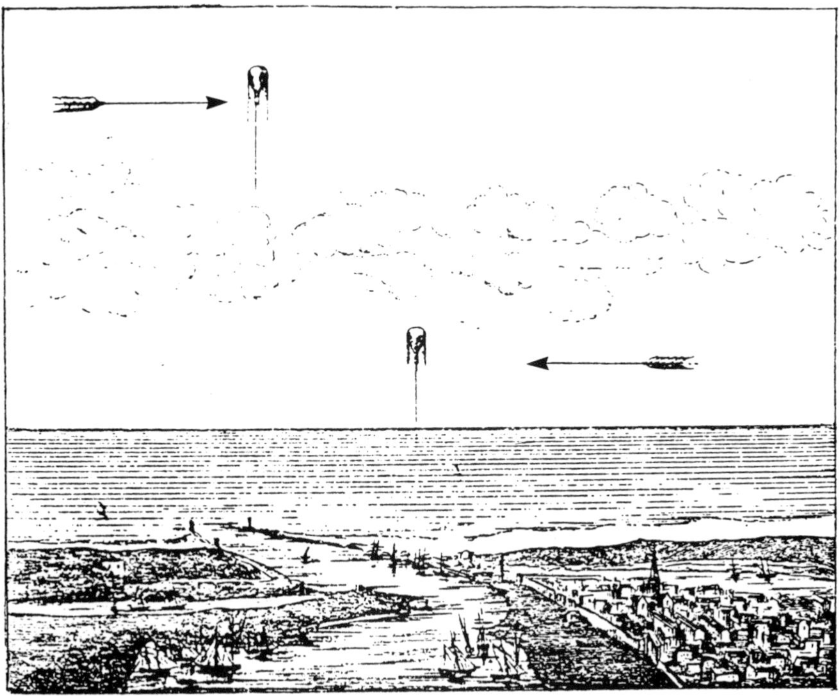

Figure 9.5. Tissandier's illustration of his use of the sea-breeze layer over Calais in 1868 to reverse the direction of travel of his balloon.

made between France and England, they succeeded in crossing the Channel. In addition, one of the most interesting was on 6 June 1883 when they succeeded in moving three times from the upper to the lower current, this verifying the accepted theory of the sea-breeze circulation (Eloy & Lhoste, 1883). Their map of this adventure is shown in figure 9.6.

Some other interesting ascents in the sea breeze, also made in France, were reported later, from Toulon ('Notizen', 1894). On 16 October 1893 the airmen Louis Godard and Jacques Courty went up at 1120 h in their balloon 'Admiral Avellan'. They started in a light breeze and fairly calm weather; at first a south-west wind carried them inland in the direction of Fort Faron. At the height of 500 m, which was reached above the Fort, the balloon began to turn towards the south, and with an ascent of 1050 m began to travel, exactly southwards in the direction of the sea. After passing the coast 1430 m above the sea the airmen travelled from the north-east above the harbour to the final protective peninsula. They had noticed from the balloon that the flags beneath were still indicating south-west. After they had passed the peninsula they let themselves down over the open sea into the lower airstream and met the desired wind at a height of 300 m. Accordingly, the balloon slid back in the exact direction of Toulon.

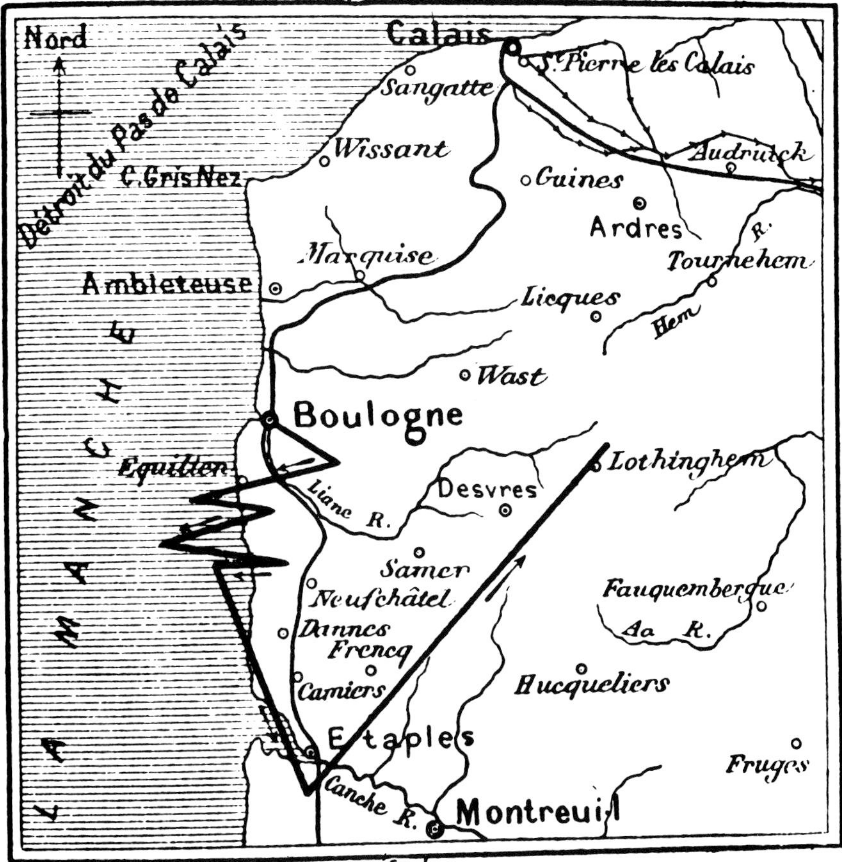

Figure 9.6. Balloon flight by Eloy and Lhoste from Boulogne in 1883 in which they moved three times between the two layers of wind.

In the second journey, two days later, the balloon rose at 1620 h and at 400 m was in an easterly airstream. At 600 m it was suddenly carried southwards towards the sea. The airmen then let themselves descend and found once more at 160 m above the sea the easterly current which brought them back above the land.

9.22 Ballooning today

Now, over a hundred years later, more balloonists than ever are taking to the air due to the hot-air revival in which modern technology has been applied to the eighteenth-century invention (Wirth & Young, 1980). The birth of the modern hot-air balloon, made of polyurethane-coated nylon and powered by a propane burner, dates from the early 1960s. Safe and reliable examples can now be purchased in many countries (Cameron, 1980).

The sea breeze, as it blows inland from coastal sites at strengths of 5 m s^{-1} up to heights of 200 to 300 metres, provides a choice of wind direction at low levels, with possible 180 degree turns. An important factor in ballooning is the possible presence of sea-breeze fronts, with their strong localised upcurrents. The probable airflow near a sea-breeze front late in the afternoon or evening is shown in figure 9.7, by which time the front may have penetrated as far as 40 or 50 km inland from the coast. The presence of the front is usually visible from the air by the distinctive form of the clouds. There may still be cumulus on the landward side of the front, which is often marked by deep clouds with a lower cloud-base. These ragged, low, 'curtain clouds' have already been illustrated (figure 9.2). As the winds converge slowly towards the front from both sides, it should theoretically be possible for a balloon to be drawn towards the front. If this seems about to happen, it is better to land before you get to it.

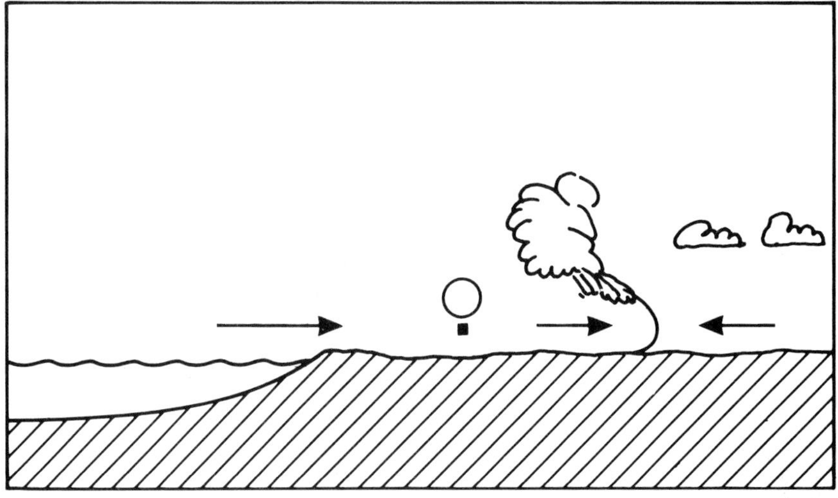

Figure 9.7. Ballooning near the sea-breeze front. The sea-air is slowly approaching the front, which the balloonist wishes to avoid.

9.3 Sailing

The weather problems of a dinghy sailor are very different from those of a deep-sea yachtsman. The latter, who may be sailing for many hours or days, needs the kind of information given in the shipping forecasts, with its pattern of pressure systems and fronts. The small boat sailor's success, and even more that of the wind-surfer, depends on being able to appreciate changes on a time scale of only a few minutes and so the understanding of local winds is of vital importance. Because so much sailing is done near coasts the sea breeze is the most important of the local winds and it is important to understand how and when it will develop and the nature of the winds related to sea-breeze fronts.

9.31 Sea-breeze sailing on a straight uniform coast

If the problems of sea-breeze sailing near a straight uniform coast are clearly understood, then it is possible to add the modifications introduced by bays and headlands and also by local topography.

In completely calm conditions the sea breeze is believed to start to blow close to the coast and gradually to extend its influence both seaward and landward at the same rate. In a flat calm this spread occurs at roughly 1 m s^{-1}, or perhaps less, and from a point out at sea it may be possible to see small disturbances in the smooth surface of the sea as the breeze gradually approaches.

Typical examples of the onset and subsequent history of the sea breeze during two days near the south coast of England are shown in figure 9.8, recorded at Thorney Island. The case in figure 9.8(*a*) was a day which started and ended with a complete calm. Soon after 0800 h some gentle suspicions of sea breeze were apparent, but not until just before 0900 h was it strong enough to record any wind strength. Soon after the middle of the day the breeze was blowing at about 5 m s^{-1} from the South, but by 1500 h it had begun to die down and by 2200 h was very weak indeed. Complete calm had returned by 2200 h. Very little veering in wind direction occurred as reported from some other sites. The only shift was a sudden one of about 20° at 1800 h, accompanied by a slight increase in wind speed.

In the case shown in figure 9.8(*b*) there was an early gradient wind blowing towards the sea from the north-east at 2 or 3 m s^{-1}. A sudden shift to the south occurred when the sea-breeze front arrived at about midday, once again the sea breeze blew from a steady direction, reducing later in the afternoon and dying completely at 1900 h. The gradient wind returned about two hours later, with light fitful gusts from its former direction.

When the synoptic wind is onshore it is difficult to detect a change as an additional sea-breeze component is added to it. The days of most interest are those in which there is a gradient wind blowing from the land towards the sea. On days of either light or offshore winds the sailing problems may be presented as four questions:

1. WILL THERE BE A SEA BREEZE?

The sea-breeze forecasting techniques discussed in Chapter 4 used the idea of striking a balance between the expected temperature rise and the forecasted

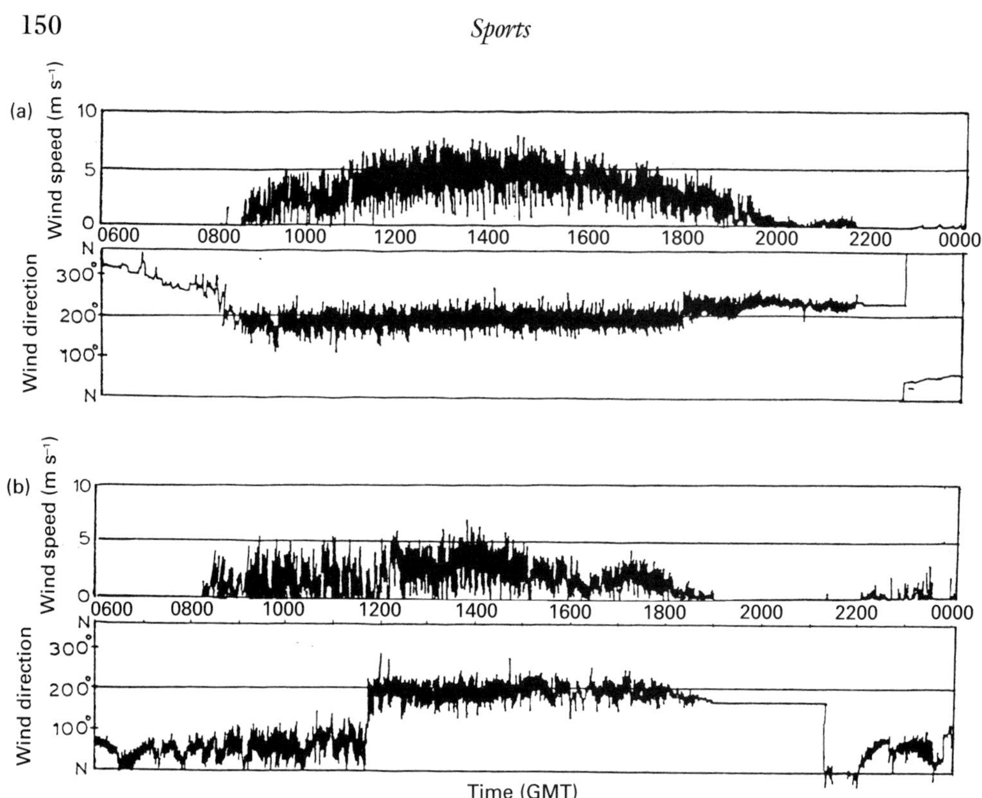

Figure 9.8. Two characteristic ways for the sea-breeze to arrive at a coastal site (Thorney Island in south England). (*a*) On a calm day. Typical early onset of a sea breeze, 2 June 1959. (*b*) On a day with light offshore wind, when a front formed at about midday. Normal onset against the wind, 13 June 1959.

wind strength. This should give a fair idea whether a sea breeze is likely to blow or not.

2. WHAT TIME WILL THE SEA BREEZE START TO BLOW?

If the time of onset of the sea breeze is plotted against the wind speed a reasonably close relationship is obtained, as shown in figure 9.9. The graph shown here is from data obtained at Thorney Island in the south of England (Watts, 1965), where it was found that on 70% of the occasions the actual times of onset were within an hour of the time given by the graph. Similar graphs could be plotted for any sailing site when enough data have been collected. The winds for the Thorney Island plot were obtained at 1000 m from radar ascents at Larkhill or Crawley, many kilometres distant. It was thought that if winds could be measured at Thorney Island and nearer the time of onset a much better fit would result. A surprising result was that the magnitude of the excess temperature seemed to have no bearing on the time of onset of the sea breeze.

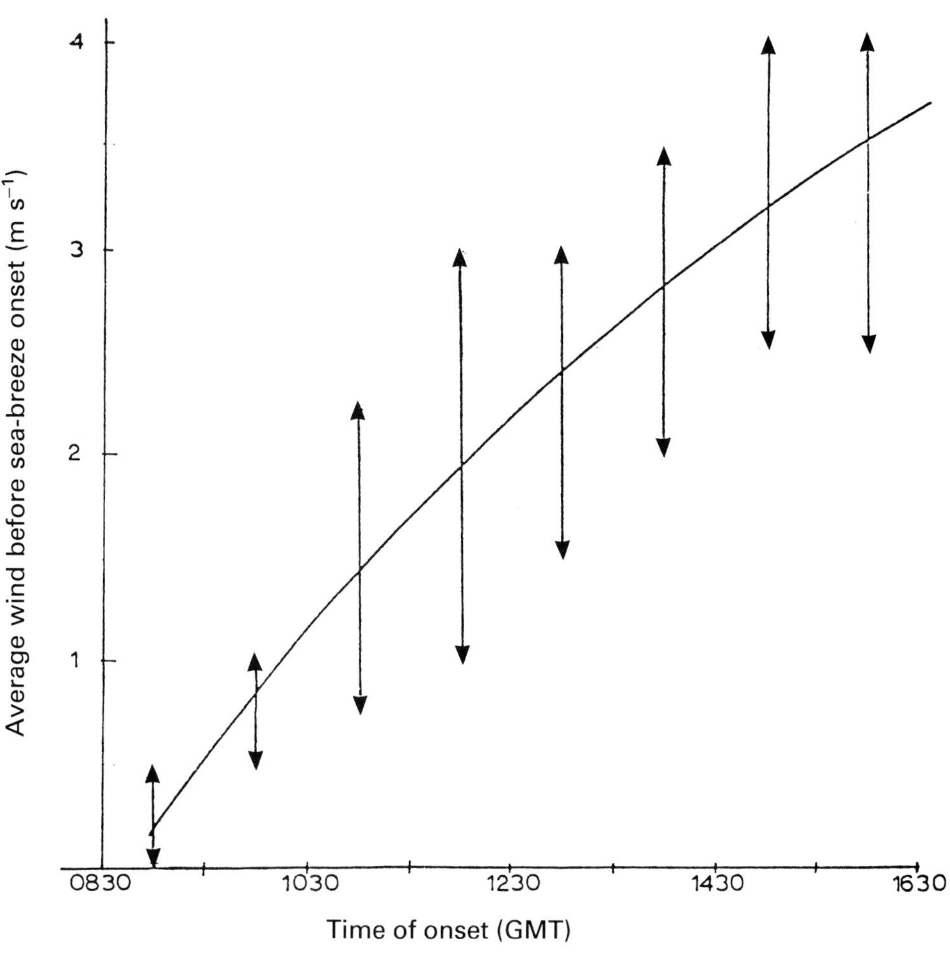

Figure 9.9. When will the sea breeze arrive? A forecast graph for the time of onset when the wind is offshore. The vertical lines show the range of wind speeds which have preceded actual onsets at the periods centred at the hour. (From Watts, 1965.)

3. HOW WILL THE SEA BREEZE CHANGE DURING THE DAY?

The changes in the sea breeze which occur during the day at any site can be summarised in a hodograph from which the direction and strength can be read off at any time of day. An example is shown in figure 9.10; this shows, among other things, that the maximum strength was just over 3 m s^{-1} and occurred at 1400 h. The forms of hodographs vary considerably at sites where they have been plotted. Among other reasons, departures from an ideal straight uniform coast affect the form of the hodograph, which may then appear very different from the example given here.

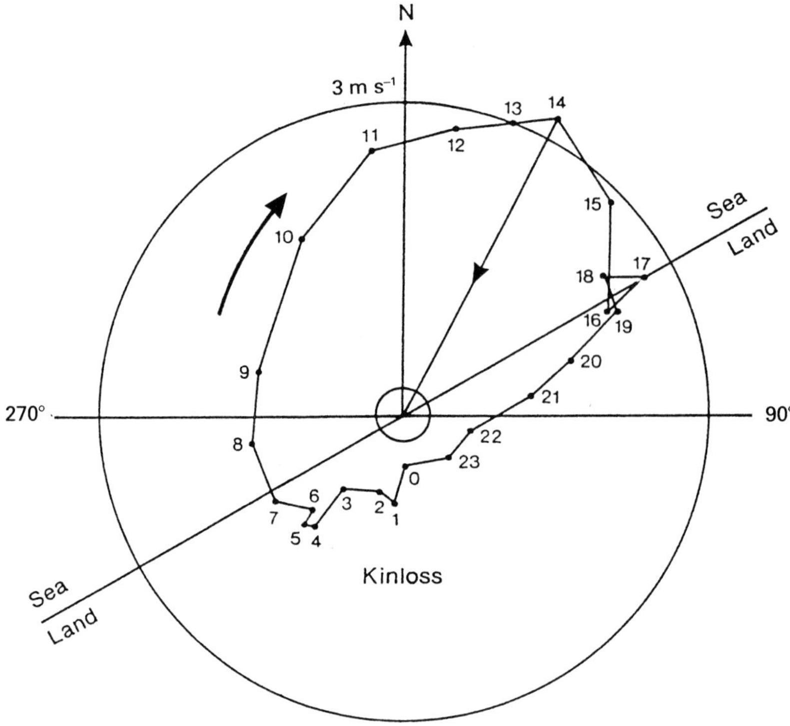

Figure 9.10. Sea-breeze change during the day displayed in a hodograph (in this case with the winds pointing inwards). The maximum wind strength was just over 3 m s^{-1} at 1400 h. The sea breeze dies between 1600 and 1900 h.

4. WHEN WILL THE SEA BREEZE DIE?

This is by no means a trivial question, as it is important to make a plan with the end of the sea breeze in mind to avoid being becalmed at an inconvenient time and place. The hodograph in figure 9.10 suggests between 1600 h and 1900 h for the death of the sea breeze.

9.32 Effects of bays and headlands

The development of sea breezes along a coastline which includes marked peninsulas and bays is more complex. The winds show a stronger sea breeze at a headland, with a convergence zone forming inland. A bay, on the other hand, produces lighter sea breeze and an area of divergence. The converging effect of such a headland is apparent in figure 9.11, taken at midday at 10 000 m near the Cherbourg peninsula on the north coast of France. The beaches on the two

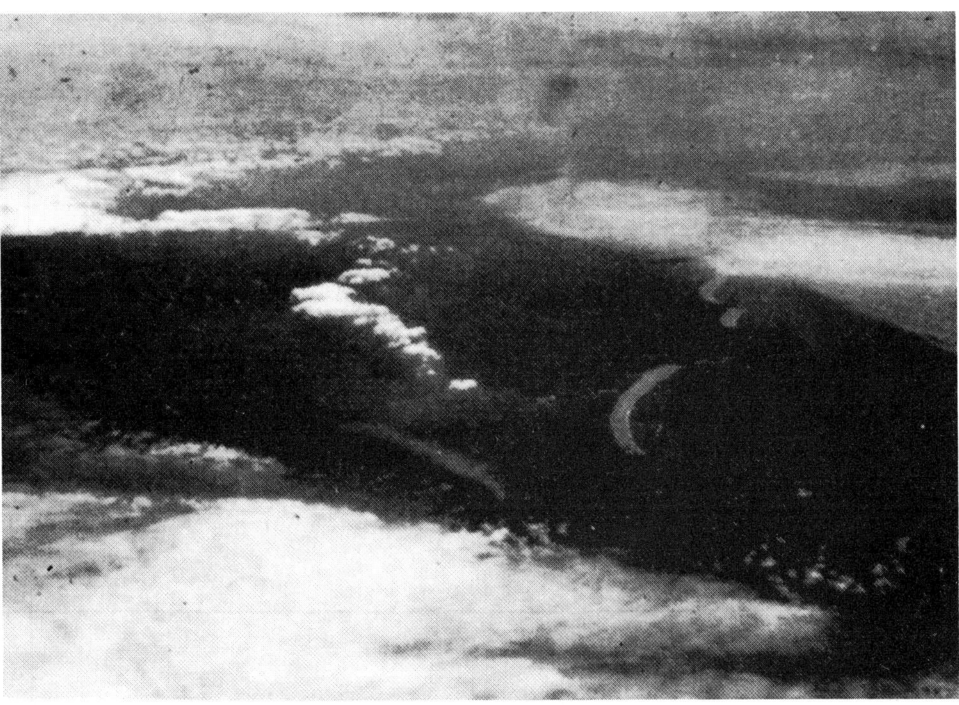

Figure 9.11. The Cherbourg peninsula, looking south from a height of 10 000 m. The beaches can be seen on the two sides, together with the line of cumulus clouds formed by the two converging sea breezes.

sides of the headland show up clearly and a line of cumulus clouds can be seen in the convergence zone formed by the sea breezes blowing in different directions around the peninsula.

Although in the early stages the sea breeze blows at right angles to the coastline, later during the day the directions of all sea breezes are likely to show a continuous change as they become directed towards the main land mass further inland. For example, in this Cherbourg case the sea breeze on the coast at the left would show an anticlockwise rotation and that at the right a rotation in a clockwise sense.

9.33 Effects of inland mountains

Wind shifts in addition to those produced by irregular coastlines may be produced by the presence of mountains at some points inland. The heating cycle during the day will produce a sequence of mountain and valley winds whose effect will be to shift the sea-breeze direction later in the day towards the highest sun-lit ground.

9.34 The sea breeze around the British Isles

The Solent may be described as the centre of south-coast sailing in Britain, and the map in figure 9.12 shows the layout between the Isle of Wight and the south coast of England. The details of the wind patterns have been described in detail by a very experienced sailing meteorologist (Watts, 1965) so we will only give the bare details here. The map shows the likely low-level sea breeze, as the island forms a barrier to these winds. Later in the day, the deeper sea breeze can flow right across the island but even so a sea breeze in the Solent is bound to have a flow round the island in its lower levels, where the sailing is done. On days when a sea breeze has been established from a wind with an easterly component blowing up the Solent it is possible that during the afternoon, due to the Earth's rotation, the sea breeze may veer during the day and blow up the West Solent from a westerly direction.

Moving further eastward along the coast the sea breezes are very similar, although the high land at Beachy Head may present an obstacle. Further east still, a strong sea breeze may arise over the Thames Estuary, from the wrong direction!

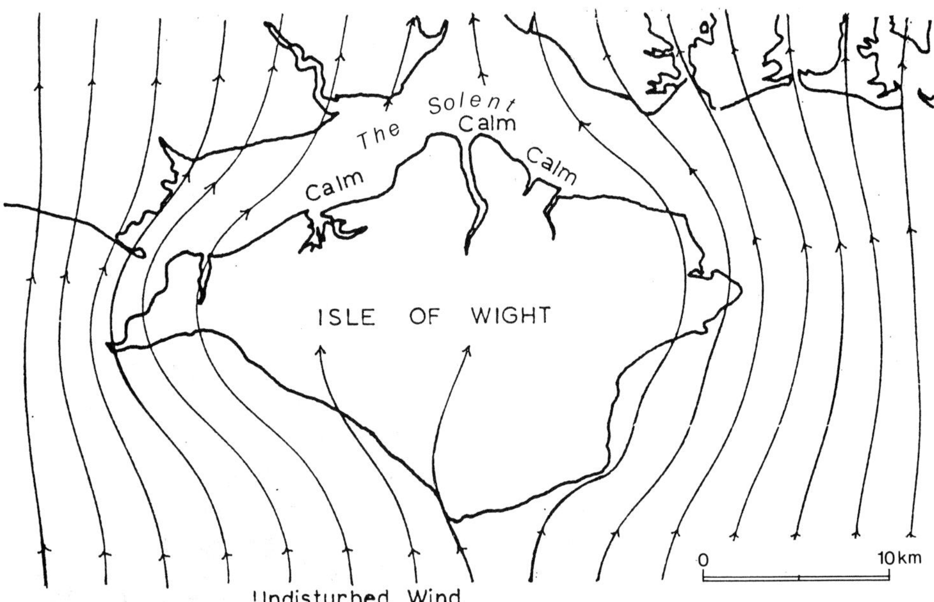

Figure 9.12. Possible airflow on a sea-breeze day across the Isle of Wight and the Solent in southern England. Later in the day the sea breeze may blow right across the island. (After Watts, 1965.)

Further north, the sea breeze blows into the Wash; when travelling from Cambridge to Peterborough, sea-breeze fronts can sometimes be seen lying parallel to the coast of the Wash.

A good sailing breeze is found at the mouth of the Tyne, but the strongest sea breezes are in Scotland, at Montrose, where the high mountains within 30 km of the coast act so as to intensify the wind.

10

Technology: field measurements of the sea breeze

10.1 Near the ground

10.11 Wind measurements: direction

The fundamental measurements of the sea breeze are of its direction and strength. The simple wind-vane to show the direction has a very long history; it has been made to represent many different objects, for example the outline of a ship or perhaps a fish will do very well to indicate the change in direction at the onset of the sea breeze, but for detailed measurements of rapid changes and of turbulence, more precise forms of wind-vane have been developed.

Recording the changes in direction with time has been done by both mechanical and electronic methods. For mechanical recording, some kind of cam may be used which rotates with the axis of the vane. Sometimes, two separate pens are employed in tracing the wind direction to avoid losing some of the record when the North reading is spread over the top and bottom of the chart. In the ingenious Baxendell recorder the paper was joined top and bottom and rolled onto a drum which rotates with the vane. The pen moved along the length of this cylinder during the day. The 'Robinson–Beckley anemograph' used a long-pitch screw to make a trace on a drum of paper rotated by clockwork. I used a modification of this design, using pressure-sensitive paper, for several years to trace the advance of the sea breeze in southern England and later in East Anglia between the coast and Cambridge. Figure 10.1 shows this instrument, with the recording drum hinged open to show the screw mounted on the axis of the wind vane. We also designed another simple instrument using an electric potentiometer, where the voltage changed with direction.

Modern instruments often employ continuous remote recording, sometimes with two self-synchronous motors.

Figure 10.1. A wind recorder for sea-breeze observations. The recording drum has been hinged away to show the screw mounted on the axis of the wind vane. The width of the recording tape is 6 cm.

10.12 Wind-speed measurements: Anemometers

1. Cup anemometer. In this simple instrument three or four cups are mounted symmetrically on a vertical axis. The force on the concave sides is more than that on the convex faces, so the cups rotate at a speed which depends on the wind speed but is independent of its direction. In some simple instruments the cups are replaced by a small propeller, but this needs to be directed into the wind.

2. Pressure tube. The open end of a tube points into the wind and the air

entering it is brought to rest. The value of the increased pressure in the tube increases with wind speed; in the famous Dynes anemometer a float in a tank of liquid is designed to make the pressure proportional to the speed.

3. Pressure plate. A plate is hung from a horizontal axis normal to the wind. The inclination of the plate to the vertical increases with the wind speed and is controlled either by gravity or by a spring. An instrument of this type was made by Hooke in 1667 and a simple version will be familiar to pilots of the Tiger Moth aircraft, in which one is mounted on a strut as a back-up air-speed indicator.

4. Hot-wire type. These use the fact that the electrical resistance of a hot wire changes with temperature. A platinum wire is heated electrically and since the current needed to maintain a constant temperature increases with wind speed, changes in this current can be recorded.

5. Sonic anemometer. This consists of a frame across which two sounds of the same constant pitch are beamed at right angles to each other. Microphones measure the pitch of each sound received at the opposite side of the frame and from the doppler shifts of the received sounds both the strength and direction of the wind are measured.

Both the hot-wire and sonic anemometers are chiefly used to examine turbulent patterns in the wind flow, but for most general purposes a simple form of cup anemometer or propeller is generally used.

10.13 Temperature

Since any property which changes with temperature can be used as the basis of a thermometer, many different techniques are used (Marshall & Woodward, 1985). There is a wide range available for many purposes, but for sea-breeze measurements we only need to measure air temperatures between 5 and 35 °C.

Suitable thermometers can be classified as follows:

Mercury-in-glass
Bimetallic
Electrical thermometers: Platinum resistance
Thermocouple
Thermistor

Mercury-in-glass thermometers are cheap and relatively robust but have several disadvantages for field work. The large mass leads to a response time in air of several minutes unless a fast current of air is forced past the bulb. The principle of operation does not allow any simple form of electrical read-out for input to a data-logger.

Bimetallic thermometers use a strip made of two different metals in which the differential expansion of the metals bends a bar which can move a pointer or recording pen. This makes a very robust instrument for field work, but the response time is slow, several minutes, and no simple electrical read-out is available.

10.14 Electrical thermometers

Several electrical properties of materials change with temperature, and electrical thermometers of various sorts have become very popular in field work due to the ease of feeding their results directly into a data-logger.

Platinum resistance thermometers use the property that the resistance of platinum is nearly proportional to temperature. The most accurate instruments employ a Wheatstone bridge circuit to reduce the current through the sensor wire and hence any self-heating effect. Most commercial instruments have a resistance of 100 ohms at 0 °C and the response times are as small as 4 seconds using the latest thin-film elements. An accuracy of better than half a degree can be obtained.

Thermocouples work on the principle that if two wires of different metals are joined at each end and the two junctions are at different temperatures then a current will flow in the two wires; the strength of the current depends on the temperature difference. The sensor element is the metal junction which can be made of low thermal capacity leading to a short response time, but the second junction has to be kept at a controlled temperature, so these instruments are generally better suited for laboratory rather than field work.

Thermistors, consisting of sensors made of semi-conducting metal oxide mixture, have many advantages which make them suitable for field work. The change in resistance is large, but is not directly proportional to temperature difference. However, many methods have been developed to overcome this difficulty. Thermistors are cheap and made in many forms: one of the most useful is made of glass or epoxy-coated beads of about one millimeter diameter. These may have a response time as short as one second.

10.15 Exposure of thermometers

A thermometer receives heat from the air by conduction, convection and radiation. Conduction only plays a small part and, since the radiation from the air itself is negligible, convection must be the main process by which the heat is transferred.

If a thermometer is to record the true temperature of the air it needs to be surrounded by some kind of screening, the main purpose of which is to protect it from radiation coming from the surroundings. Many kinds of screen have been used for this purpose, the most familiar being the Stevenson screen with louvred walls made large enough to contain thermometers and perhaps some larger recording instruments. For some of the smaller thermistor instruments a different form has been developed consisting of a series of conical metal shields painted with a high reflectance white epoxy.

10.16 Humidity

The humidity is a measure of water vapour pressure present in the atmosphere and is very important in sea-breeze studies. Many possible methods exist for estimating this but it is difficult to measure accurately. Three different ways in which it is usually expressed are listed below; given the vapour pressure and the temperature each of these humidity measures can be calculated.

1. *Dew point temperature* is defined as the temperature at which the air would become saturated if cooled at constant pressure. The dew point of an air sample can only be changed by evaporation or condensation, so it is a useful property in measurements of different air masses associated with the sea breeze, such as may occur at sea-breeze fronts.
2. *Humidity mixing ratio*, q, is defined as the amount of water vapour in grams mixed with one kilogram of dry air. Common values are in the order of 5 to 30 g kg^{-1}.
3. *Relative humidity*, usually expressed as a percentage, is the ratio of the actual value of vapour pressure to the saturated vapour pressure at air temperature. Unlike the dew-point, the relative humidity depends not only on the vapour pressure but also on the temperature.

10.17 Humidity measurement: hygrometers and psychrometers

Some types of instrument are suitable for measuring humidity in field work; there are many others more appropriate for use in the laboratory.

MECHANICAL (HAIR) HYGROMETERS

Some materials increase in length as the relative humidity increases, one example being human hair. Instruments are available in which the changes in length with humidity of a bundle of hairs are increased by leverage and either displayed on a dial or recorded by pen on a chart. Although not capable of great accuracy these simple cheap instruments are still in wide use.

WET- AND DRY-BULB HYGROMETERS (PSYCHROMETERS)

The instrument consists of two thermometers, one of which has its bulb permanently moist. When air blows across the instrument, the wet-bulb thermometer reads lower than the dry-bulb, and the depression give a measure of the humidity. These are probably still the most frequently used instruments for measuring humidity.

It has been found that, for a given humidity, the wet-bulb depression increases with the draught speed. However, this effect is important only with wind speeds less than 3 m s^{-1}, increasing the draught further has no effect. As long as the wind speed outside the screen at least equals this the readings are reliable. For greater accuracy in light wind conditions, an instrument which maintains a forced draught of at least 3 m s^{-1} can be used. The simplest of these is a whirling wet and dry bulb hygrometer, useful in field work when not using a screen. A more complex instrument is the Assman psychrometer which draws a forced draught across the two thermometers mounted within a radiation shield.

HYGROSCOPIC SENSORS FOR ELECTRICAL RECORDING

Sensors exist whose resistance depends on humidity, but very effective instruments developed recently employ changes in capacitance of a thin film as it absorbs or releases water vapour. The principle of the Vaisala dielectric polymer sensor is shown in figure 10.2. The sensor element is small (approximately

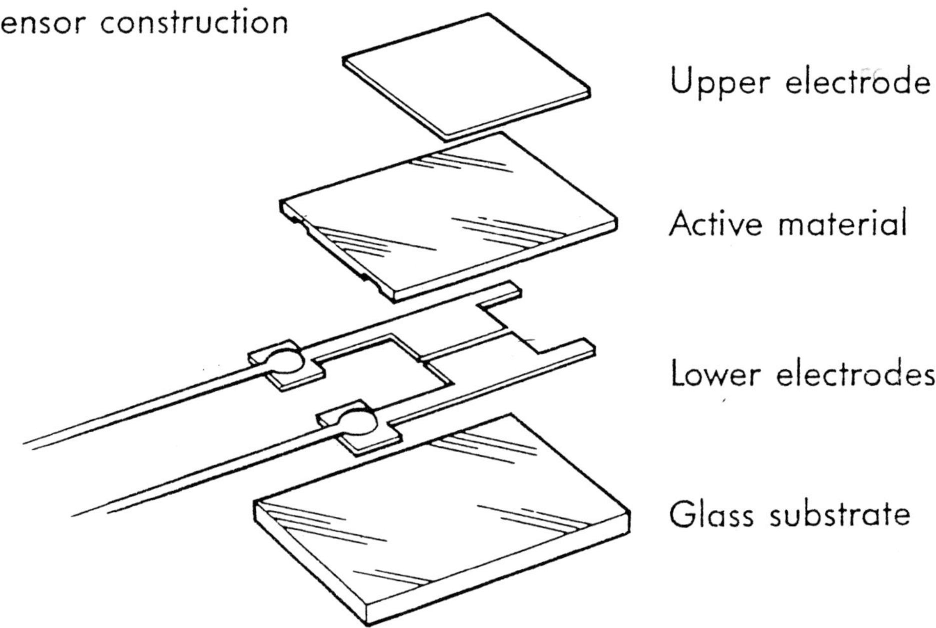

Figure 10.2. Principle of the Vaisala dielectric polymer sensor for measurements of humidity.

4 × 6 × 1 mm) and the upper electrode is so thin that water can diffuse through it rapidly and response times of the order of one second are available.

SPECTROSCOPIC METHODS FOR HUMIDITY

These are complex and expensive and mostly more suitable for laboratory measurements, but a commercial instrument was available in the 1970s which measured water vapour pressure by the absorption of the Lyman alpha line in the ultra-violet spectrum. Although this instrument needs careful calibration before and after each flight, its response time is less than a second. I found this instrument invaluable for airborne measurements at sea-breeze fronts using a light aircraft.

10.18 An automatic weather station

Some of the instruments we have described are mounted in the simple automatic weather station illustrated in figure 10.3. The wind cups and wind vane can be seen at the ends of the cross beam, which has a radiation recorder at its centre.

Figure 10.3. A simple electric recording station, consisting of a wind cup and wind vane at the ends of a cross beam. Temperature and humidity sensors are mounted at an angle, under a radiation shield. (Courtesy of Delta-T Devices.)

Figure 10.4. Three time-lapse frames taken at intervals of two minutes from a film of curtain clouds forming at a sea-breeze front.

The radiation shield for the temperature and humidity sensors is mounted at an angle to the main support. The data-logger is shown mounted on the mast, which is supported by guy cables.

10.19 Time-lapse cloud photography

Speeded-up motion photography has become familiar from its use in showing developing plants. Striking pictures can be made to show the unfolding of a flower-bud and the spreading petals.

The development of clouds is also difficult to perceive at normal rates, but can be made clearer by speeding up the motion. This is done by taking pictures much slower than the rate of 25 frames per second at which they are usually projected. The study of the development of cumulus clouds is a very rewarding subject, and I have taken many hours of film of cloud forms at sea-breeze fronts in order to understand what is happening there (Simpson, 1968).

Time-lapse film is capable of useful measurements of the air currents involved in a developing cloud; however, one of the difficulties in making a useful time-lapse cloud film is that the whole cloud often moves out of frame before much development is visible. The easiest subjects are wave clouds and cumulus clouds on days of very light winds. The comparatively slow advance of a sea-breeze front makes it another very suitable subject; for example, the three time-lapse frames in figure 10.4 show the rapid development of curtain clouds with a low cloud-base beneath a sea-breeze front. These were taken at Lasham in southern England in the late afternoon, looking south towards the slowly approaching front. The time interval between these pictures is two minutes; they were selected from frames taken at intervals of three seconds.

10.2 Airborne measurements

10.21 Pilot balloons

Watching the behaviour of a small buoyant balloon as it ascends through the sea breeze is a simple and effective way of seeing the direction of the breeze and the height through which it extends.

Useful measurements can be made using balloons with standard rates of ascent, and measuring the direction by theodolite. Near a sea-breeze front, however, where rising air may be expected, two theodolites are needed, or, more simply, a 'tail' from the balloon is measured to give the distance from the observer. For sea-breeze work a balloon weighing 30 g is suitable, inflated by helium to a diameter of one metre. The tail, suspended 20 m beneath the balloon can conveniently be another balloon, inflated with air. The elevation, azimuth and graticule length of the tail can be recorded every 30 seconds and hence the required wind speed and direction calculated to a height of one or two thousand metres.

10.22 Radio-sondes and rawinsondes

A radio-sonde consists of a small instrumental package which is carried aloft by a large balloon and which transmits instrument readings of temperature, humidity and pressure at regular intervals, usually of 15 s, to a simple receiving station on the ground. The 100 g balloons are usually inflated to give about 3 m s^{-1} rise, yielding a vertical resolution of 50 m.

A rawinsonde uses in addition radar to track it and hence measure the wind. It may also receive time differences from the Loran navigation system to produce profiles of horizontal wind components.

10.23 Tetroons

A tetroon is a balloon made in the form of a tetrahedron and intended to maintain almost constant volume. As it rises, instead of expanding as would a normal spherical balloon, and thus keeping a buoyance difference from its surroundings, it can be designed to reach equilibrium at some desired height. As described in Chapter 6, tetroons have done useful work near sea-breeze fronts, where they have been carried up by the rising currents and in some cases have made complete circulations.

10.24 Measurements from aircraft

Many of the types of instruments described above have been carried aloft in all sorts of aircraft to make 'on the spot' recordings. The results can be recorded by data-loggers on computers.

Some comparatively inexpensive work at sea-breeze fronts has been done using gliders and especially 'motor-gliders'. These have many of the characteristics of gliders but the use of power widens their scope.

A motor-glider has been also used as a sensor of vertical air motions in meteorological research in thermals and at sea-breeze fronts. Experiments with a Scheibe Motor-Falke (Mansfield, Milford & Purdie, 1974) have shown how it is possible to do without expensive equipment such as inertial platforms and to use the aircraft as nearly a passive sensor as possible.

10.3 Remote sensing: radar

As its name suggests, radar was developed for the purpose of radio detection and ranging of aircraft. After World War 2 many radars were employed for probing the atmosphere and various unidentified echoes were obtained which came to be known as 'angels'. I recall joining a friend using a radar on a trailer in southern England in about 1960 on a hunt for angels. He gave me the azimuth and range of moving echoes as they came from a target onto which he was locked and I plotted them on the map. They were moving at roughly 15 m s^{-1}, and I remember the excitement when we plotted three successive above Lasham Church. We were unable to spot any visual targets overhead in the dark.

At the end of the 1960s many angel echoes associated with the sea breeze had been recorded; by this time the cause of most 'angels' had been explained and the name was gradually dropped. Sea-breeze fronts were identified by radar at Wallops Island on the east US coast (Atlas, 1960). No birds or other solid objects were seen and an explanation was proposed in terms of refractive index gradients formed at the sea-breeze interface.

10.31 History of radar detection of sea-breeze fronts

On 20 June 1960 a powerful research radar was being operated near Chelmsford in south-east England by Marconi (whose chief engineer at the time was Eric Eastwood, an enthusiastic ornithologist). A time-lapse movie was made of the PPI (Position plan indicator), which showed the development of a front from midday until after sunset (Eastwood & Rider, 1961). By 1730 GMT the front had moved 70 km inland, the total length of the north and south segments was nearly 200 km, and it was still advancing at 3 m s^{-1}. The results of the PPI image at this time are shown in figure 10.5.

The sea-breeze explanation of this remarkable echo was inferred from meteorological evidence. There was little general wind and almost no low cloud and the changes in wind speed and direction measured at the ground indicated the arrival of the sea breeze at weather stations during the day.

One feature of the radar pictures was the sharpness of the leading edge as compared with the trailing edge of the front. 'Angels' were detectable moving away from the line at speeds of about 15m s^{-1}. Apart from the clear solid images caused by aircraft a stream of angel echoes could be seen moving up the Thames Estuary.

The interesting question to be answered was what property of the sea-breeze front caused it to be visible on radar. No clouds or rain were present to form

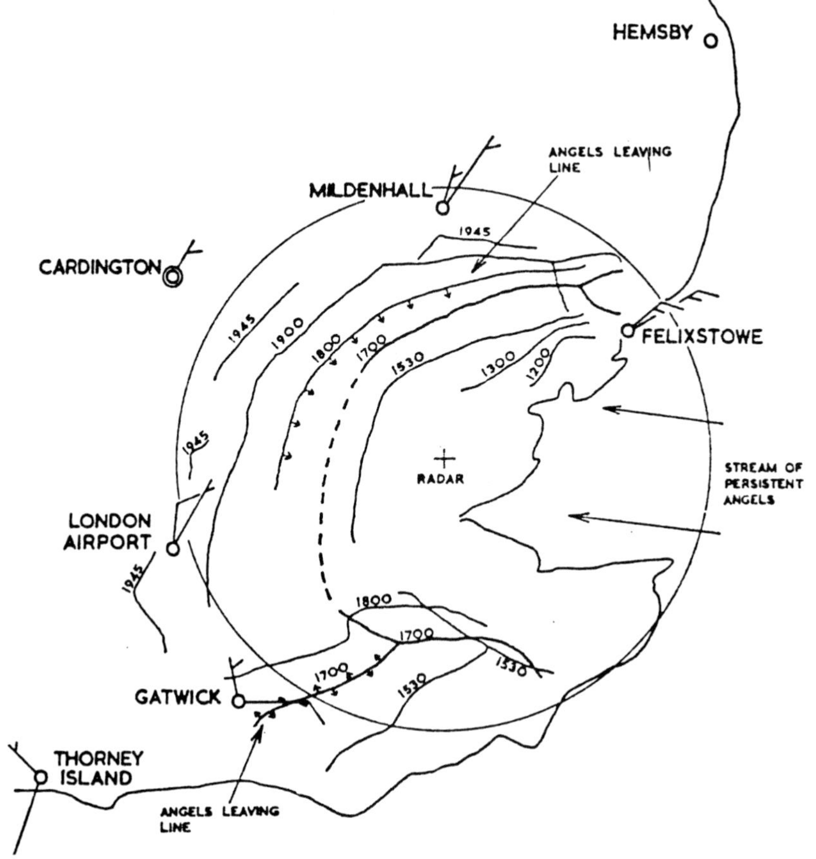

Figure 10.5. Development of radar echoes of a sea-breeze front from 1200–1945 GMT on 20 June 1960 in south-east England. The range ring is at 40 nautical miles (about 70 km). (After Eastwood & Rider, 1961.)

a radar echo and two possible explanations of the visibility of the front to radar remained.

1. Reflection from a refractive index discontinuity at the front.
2. Scattering attributed to birds.

REFLECTION FROM REFRACTIVE INDEX DISCONTINUITY

In the case of the sea-breeze incident described here, a simple theory of discontinuity between the moist, cool sea air and the dry, warm land air was first considered (Eastwood, 1967).

If the discontinuity took place within a few centimeters over the whole sheet intercepted by the radar beam the received signal was calculated to be well below the threshold of detection.

If, however, a region of mixing existed there could be many interfaces capable of reflecting radio energy and it seemed possible that a small contribution to the radar signal received could be produced by the mixing region present at the frontal interface.

SCATTERING ATTRIBUTED TO BIRDS

The above argument suggested that birds probably provided the bulk of the radar signal.

As mentioned above in the section on birds in the sea breeze (section 8.3), glider pilots often met groups of swifts soaring at sea-breeze fronts, and several experiments showed the presence of swifts at fronts seen on radar. Figure 8.14 shows one such radar display in southern England from the Marconi radar on a day in which I flew a light aircraft counting swifts at a sea-breeze front. The echoes from the aircraft and also from several groups of swifts seen at the sea-breeze front are clearly visible.

What one detects on radar depends quite dramatically on the radar resolution, its sensitivity and its target range; it may also vary with the wavelength employed. Echoes associated with sea-breeze fronts were observed during the 1960s using radars operating simultaneously at two different wavelengths of 3.2 and 10.7 cm. (Geotis, 1964). Modest concentrations of insects in the sea-breeze convergence zone were sufficient to explain the magnitude of the echoes, their wavelength dependence and the fluctuating nature of the signal.

As well as the information shown above in the familiar PPI mode (position plan indicator) some useful sea-breeze information has been displayed in the RHI mode (range height indicator). An example is given in figure 10.6, which was taken in 1966 of the sea breeze at Wallops Island, looking north. The higher of the two layers shown is at about 1000 m, so the sea-breeze front, which appears beneath them on the left, is about 300 m deep.

10.32 More recent sea-breeze investigations

Some useful records were made of sea-breeze front development in the south of England using the Chilbolton Radar, backed up with ground observations and using an instrumented motor-glider.

The large Chilbolton 25 m steerable aerial, which has many special features related to radio-wave frequency research, (Meadows, 1967) is illustrated in figure 10.7. Clear radar echoes were obtained from the area of the sea-breeze front on 4.5 and 7 June 1973; on all three days the front reached as far as 20 km inland.

The greatest number of clear photographs taken was on 5 June, and the first of these showed faint diffuse radar echoes. Random dot echoes, seen by 1400 GMT, later linked to form a line about 15 km inland; see figure 10.8.

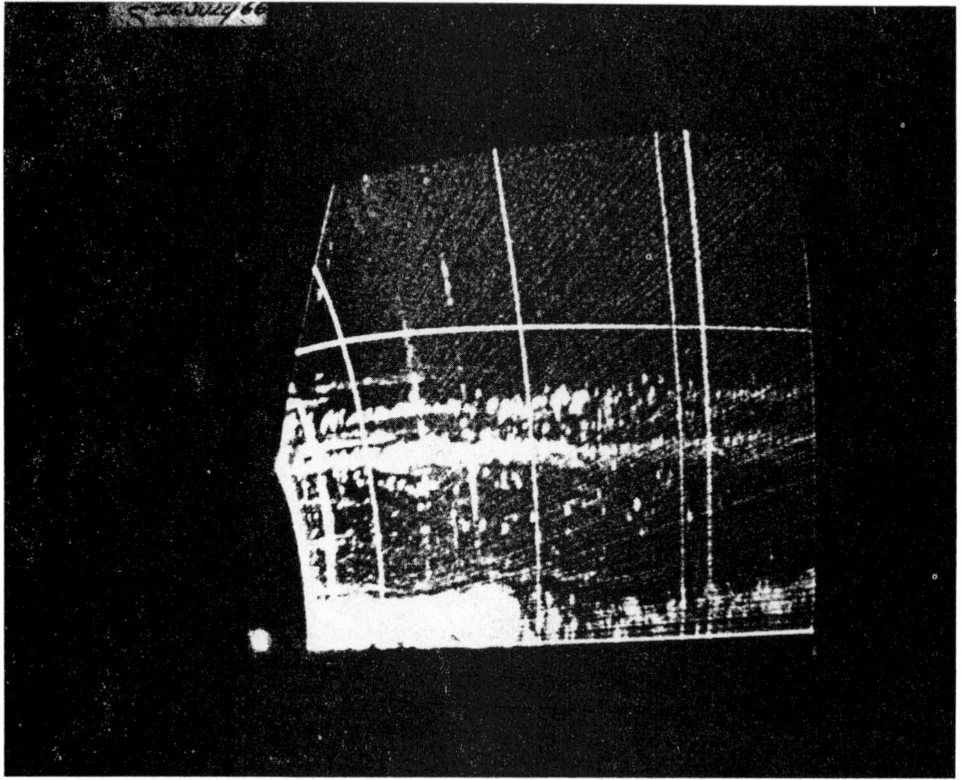

Figure 10.6. Radar display in 'range-height indicator' mode, showing a sea-breeze front at Wallops Island, looking north, on 26 July 1966. The sea breeze, seen at the bottom left, is about 300 m deep.

The diffuse radar echoes from clear air were thought most likely to be due to fluctuations in refractive index. Two almost continuous lines appeared, the upper line was in the region of wind shear, which was established from pilot balloon ascents. The lower line of echo can be associated with the maximum refractive index fluctuations at the sea-breeze front, and its position is consistent with the structure as measured from an aircraft (Simpson, Mansfield & Milford, 1977).

The wavelength of the radar used in the Chilbolton sea-breeze investigation was 'S-band'. This is one of the longer wavelengths used in 'meteorological radar'. Those usually employed are:

L-band (15 to 20 cm)
S-band 10 cm (15 to 5 cm)
C-band 5 cm (8 to 4 cm)
X-band 3 cm (4 to 2.5 cm)

Figure 10.7. A steerable radar aerial, 25 m diameter, at Chilbolton, Wilts, England.

10.33 Local airflow revealed by radar observations of insects

Radar-detected insects can be used as natural tracers of the windflow whenever the wind is strong enough to dominate their movement. Most of this information has already been discussed in Chapter 8 on insects in the sea breeze; the most important contributions from radar will only be summarised here.

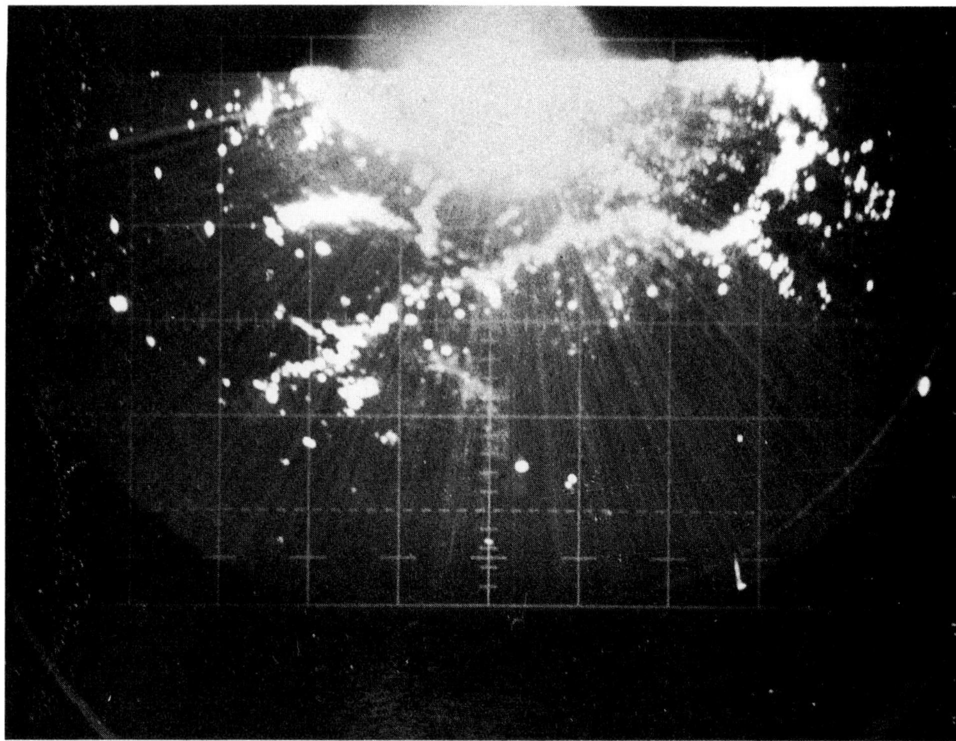

Figure 10.8. A sea-breeze front in southern England seen by Chilbolton radar, at 1707 GMT, 5 June 1973. The radius of the ring is 50 km.

A 3 cm radar has been used by the Centre for Overseas Pest Research to study the night flight of the African armyworm moths *Spodoptera exempta* (Pedgley *et al.*, 1982). For example, it was possible to describe outflows of cold air sweeping up insects to the windshift line; the observations gave a view of the process of concentration of airborne insects within a zone of convergence.

The sea-breeze front at Canberra traced by 3.5 cm radar shows the presence of many flying insects, mostly grasshoppers. Their presence has been used to show the position and progress of the fronts and to outline their leading edges (Drake, 1982).

In East Canada a four-year study in New Brunswick using radar to investigate the spruce budworm moth (*Choristoneura fumiferana*) traced the position of sea-breeze fronts (Dickison, 1990). The use of airborne doppler radar navigation and wind-finding systems led to the identification and forecasting of wind-convergence systems.

Quantitative information about the microstructure of the sea-breeze front circulation and concentrating mechanism was obtained from airborne downward-pointing radar by Schaefer (1979).

10.34 Velocity measurements of airflow at sea-breeze fronts

Doppler radars not only detect the presence of a target but can measure its speed towards or away from the radar. Modern doppler weather radars are sensitive enough to detect clear air echoes in the neighbourhood of sea-breeze fronts and a good result obtained from such techniques is shown in figure 10.9. These measurements were made at Norrkoping in Sweden and display the horizontal velocity field around a sea-breeze front advancing inland from the Baltic Sea (Andersson & Lindgren, 1992). It shows a sea breeze of strength 6 m s^{-1} and depth 600 m which was advancing against an opposing wind of about 4 m s^{-1}. The front at this time (1400 GMT) was 12 km from the coast and advancing at about 5 km hs^{-1}.

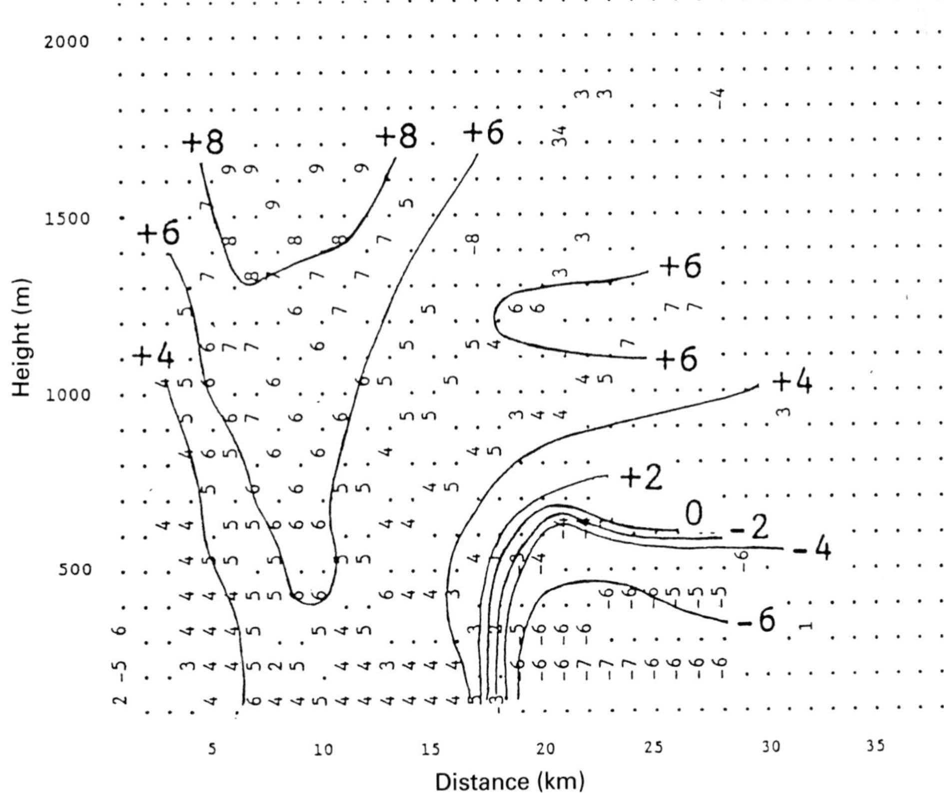

Figure 10.9. Horizontal winds (in m s^{-1}) in a cross-section of a sea-breeze front measured by doppler radar. The front was 12 km inland from the Baltic Coast, at 1400 GMT on 7 May 1990. (From Andersson & Lindgren, 1992, by permission of HMSO.)

10.35 Radar observations of land-breeze fronts

Compared with those of the sea-breeze front observations of the land-breeze front, by any method, are very scarce.

A land-breeze front is the boundary that separates the cooled dense air moving seaward from the warmer, less dense, air above the sea. When the gradient wind over the area is light and onshore, conditions may be favourable for the formation of a land breeze front. This was the case on 28 and 29 August 1969, when the 10.7 cm wavelength radar at Wallops Island, Virginia recorded the landbreeze phenomena which existed off the coast (Meyer, 1971).

The radar display in the morning of 28 August, at 1043 EST. is shown in figure 10.10. The distance from the shore was then about 24 km, and the front was lying roughly parallel to the coast, which lies from north-west to south-west. Later radar displays showed the front receding and dissipating. A similar pattern was seen on the following day.

LAND-BREEZE CONVERGENCE OVER THE ENGLISH CHANNEL

The image related to a land breeze convergence zone over the English Channel

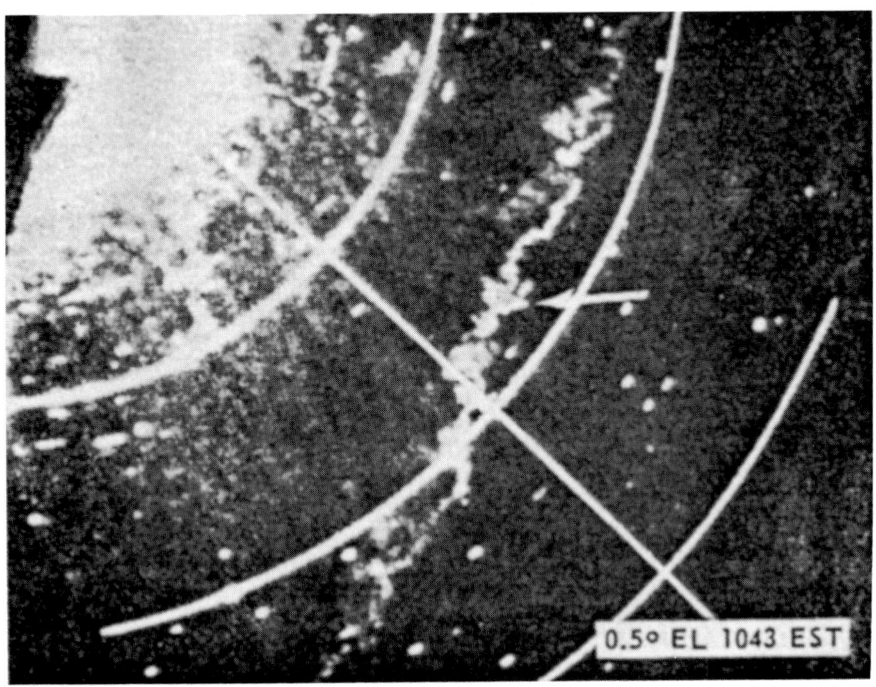

Figure 10.10. Land breeze front seen by radar 15 nautical miles out to sea near Wallops Island at 1045 EST on 28 August 1969.

shown earlier in figure 7.16 was based on rainfall intensity from the UK weather radar network (Waters, 1990). It was made at 0600 GMT on 19 October 1990 and followed a night during which the first rain cells developed over Lyme Bay at 2300 GMT, thought to have been locally enhanced by the focussing of offshore land-breeze circulations.

10.4 Remote sensing: acoustic sounding or sodar (or sonar)

It was not until 1968 that effective echo sounding was achieved in the atmosphere (McAllister *et al.*, 1969). Compared with radar, relatively simple equipment can be used and it was shown how easily the echoes could be obtained and displayed.

The absorption of the wave as it propagates through the atmosphere is large compared with radar and this sets a limit to the range through which useful echoes can be obtained. The best frequency is found to be between 1000 and 2000 Hz.

In its simplest form, sodar (as acoustic sounding is now usually named) consists of a vertical pointing loudspeaker, or array of speakers. About half a second after transmitting a burst of sound the array is switched to listening mode and the echoes received are amplified and displayed on a facsimile recorder. An arrangement in which the same array is used for transmission and reception is called a monostatic sodar. Systems in which the transmitter and receivers are separated and off-axis sound is detected are called bi- (or tri-) static sodars.

10.41 Sodar and the sea breeze

The acoustic scattering which can be detected by sodar is produced by small-scale temperature fluctuations in regions or temperature and wind gradients; strong echoes may be expected from boundaries of the sea breeze.

The development of the sea-breeze, or 'lake-breeze', layer during the day above a point near the edge of Lake Ontario is shown by monostatic sodar in figure 10.11 (Bennett & List, 1975).

The lifting of the nocturnal inversion began at 0700 EST and continued to be visible until 0900, when it disappeared. The lake-breeze front appeared at 0920 EST, producing an intensification of the lower-level signal. At 1015 EST a stable layer was formed, which continued throughout the day.

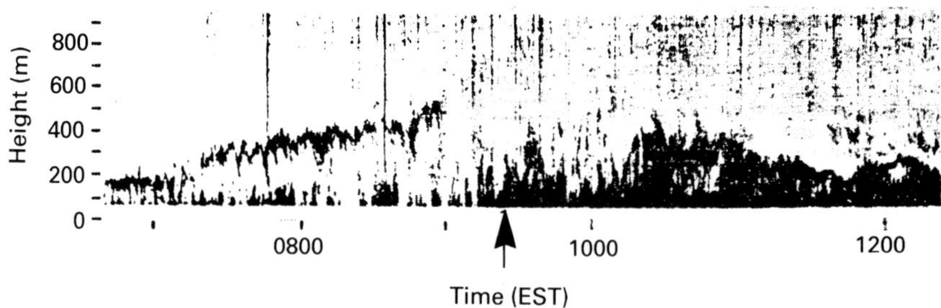

Figure 10.11. Sodar echoes for 26 April 1974 near the edge of Lake Ontario. Between 0700 and 0900 EST the nocturnal inversion lifted, and at 0920 EST the lake-breeze front arrived (arrow). (After Bennett & List, 1975.)

10.42 Doppler sodar

Tristatic doppler sodar, using three receivers, gives information of velocities in the echo field by measuring the changes in frequency detected in the set of receiving points.

Information on the microstructure of the sea-breeze front and the evolution of the sea breeze in general was obtained in 1987 during the LASBEX project (Land and Sea Breeze Experiment), Intieri *et al.*, 1990, whose equipment included tristatic doppler sodar. Doppler sodar results from Monterey Bay, in which the wind direction and strength were displayed, clearly indicated the onset of the sea-breeze front. Such sea-breeze fronts were measured by the sodar on 10 of the 14 measurement days of LASBEX.

10.5 Remote sensing: lidar

Laser radar, usually called lidar, was first used to observe aerosol distribution in the atmosphere in 1963. Lidar receives back-scattered light from airborne particulates, and does not record turbulent fluctuations in atmospheric temperature or humidity, as usually recorded by sonar and radar observations.

Lidar has been used in the atmosphere (Shimizu *et al.*, 1985) employing the plan position indication (PPI) and range height indication (RHI) modes. The structure of both the convective mixed layer and the fronts of gravity currents have been studied.

At the beginning of lidar history ruby lasers were mainly used, and later many other systems were employed.

People are protected from the laser beam by various methods. For the safety of those in nearby buildings higher than the laser output, laser firing automatically stops when the scanner points in those directions. Aircraft are protected by a radar system attached to the receiving telescope, so firing is automatically stopped if an aircraft is detected within the viewing angle of the radar.

10.51 Sea-breeze measurements by lidar

One of the earliest applications of lidar to the sea breeze was in the measurement of polluted layers near Chicago by Bennett & Oded (1967). Vertical soundings were made of particulate concentration with height during a pollution episode on 26 April 1966. The peak concentration at 130 m is at least 3000 mg m^{-3}. The data show that the inhabitants of high-rise buildings may be exposed to air of poorer quality than those who live near the ground.

10.52 Lidar and the sea-breeze front

The lidar display in figure 10.12 is the vertical cross section of a sea-breeze front, showing the distribution of aerosol concentrations (Nakane & Sasano,

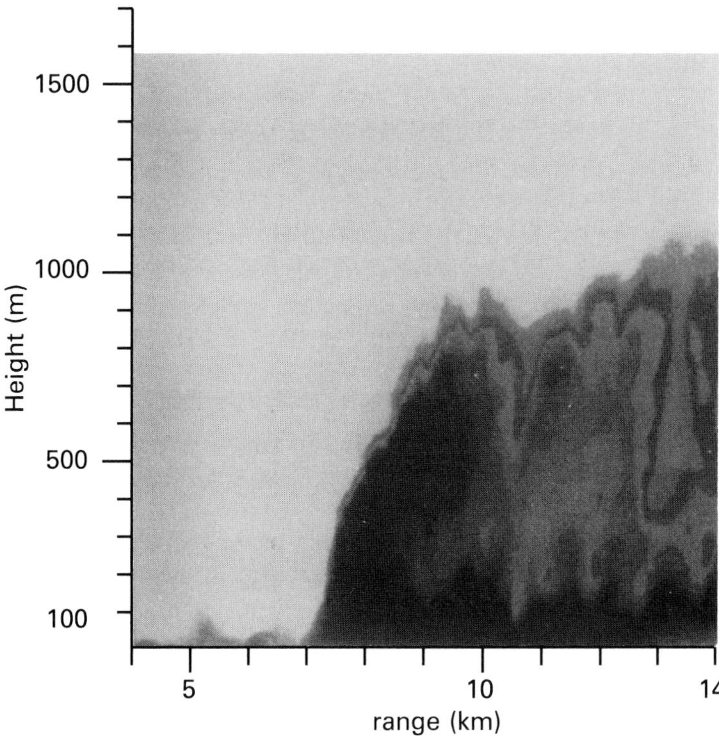

Figure 10.12. Lidar display of a sea-breeze front, 45 km inland from Tokyo. The dark regions show higher concentrations of aerosols. (Courtesy of M. Nakane.)

1986). This observation was made at 1740 hours, when the front, moving at 3.3 m s^{-1}, had reached a point 45 km inland from Tokyo, Japan. The results confirm the structure of the sea-breeze front as previously measured from aircraft flights.

10.53 Doppler lidar measurements of a sea-breeze system

September in Monterey Bay, California, is a month of strong sea breeze and minimal fog. In September 1987 doppler lidar was used in the LASBEX project, together with sodar and other sensors, to investigate the vertical and horizontal structure of the sea breeze during diurnal cycles (Intieri *et al.*, 1990).

The vertical structure of the sea breeze is revealed in the three lidar RHI scans, on 16 September 1987, in figure 10.13. At 1617 UTC a 1000m deep land breeze blew steadily from the right (west) at 7 m s^{-1} figure 10.13(*a*). The winds near the surface became lighter and at 1720 a sharp sea-breeze front passed the station. A breeze from the land was still blowing above the shallow sea breeze layer at 1900 figure 10.13(*b*). Within 4 hours of its initiation (at 2100 UTC) the sea breeze reached its maximum depth of 1000 m and wind speed of 8 m s^{-1}, figure 10.13(*c*).

10.6 Satellite imagery

Since the early days of space flight many photographs have been taken showing cloud forms close to the Earth's surface. Such pictures often show cloud patterns from which the extent of the sea-breeze development can be inferred. The value of the study of cloud patterns viewed from aeroplanes has already been mentioned in Chapter 1; however, satellite viewing makes it possible to see sea-breeze patterns covering a complete continent. For example, an early Gemini photograph showing sea-breeze clouds over the whole continent of India was shown in figure 2.13. One feature of interest is the wide cloud-free zone along a large strip parallel to the west coast where the sea-breeze is blowing.

10.61 Types of satellite observation

Meteorological satellites are of two main classes, geostationary and polar-orbiting satellites. A geostationary orbit is one at a height of 36 000 km where the period of rotation of the satellite is 24 hours and therefore its position over the Earth's surface remains constant.

A chain of geostationary satellites exists around the equator, the European

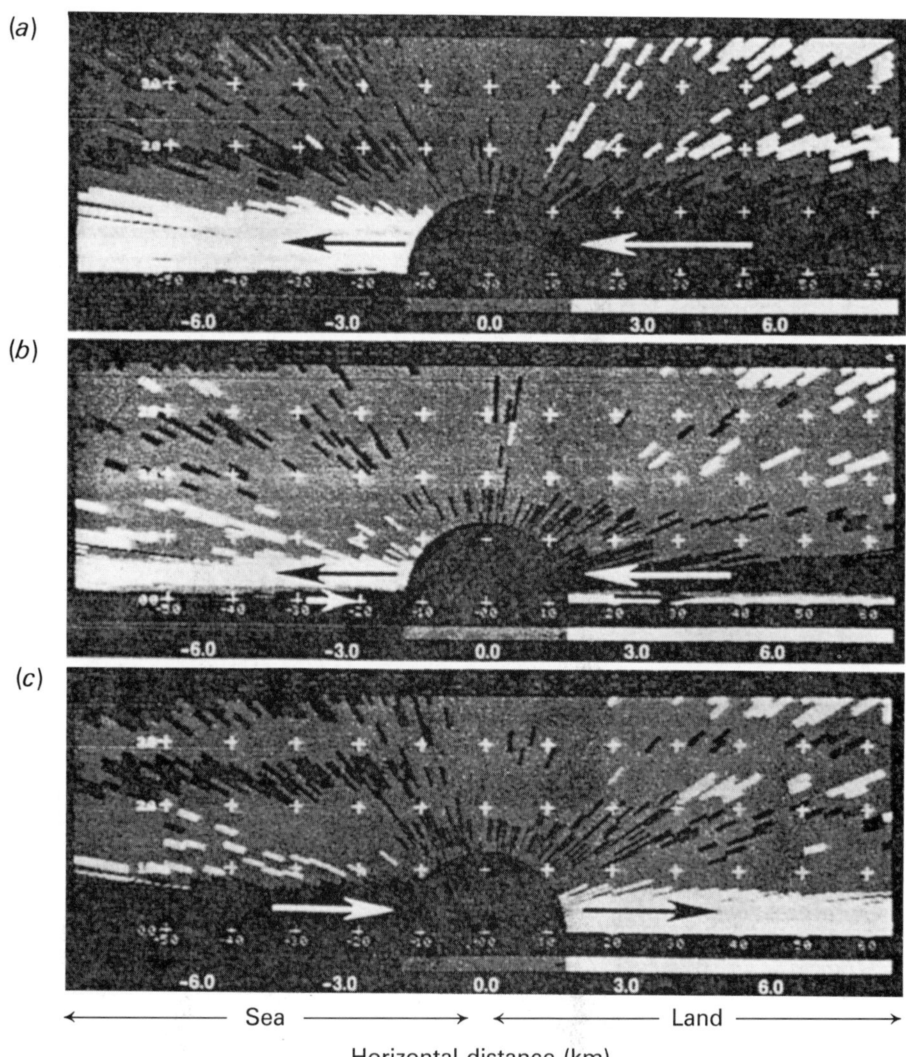

Figure 10.13. Three lidar RHI scans on 16 September 1987 at (*a*) 1617 UTC, (*b*) 1900 UTC and (*c*) 2100 UTC. The sea is on the left and the land on the right. The darker shades represent flow approaching the lidar, in the centre, and the lighter shades show flow away from the lidar. (From Intieri *et al.*, 1990.)

link of which is Meteosat-4, launched in 1989. It is located over the point where the Greenwich Meridian crosses the equator. The visual, infra-red and water vapour images are not directional, but are broadcast all over Europe for anyone who has the equipment to receive them. Until recently this has meant almost exclusively government meteorological services and the military. But now equipment is available, using a dish and personal computer, to download and construct animated sequences.

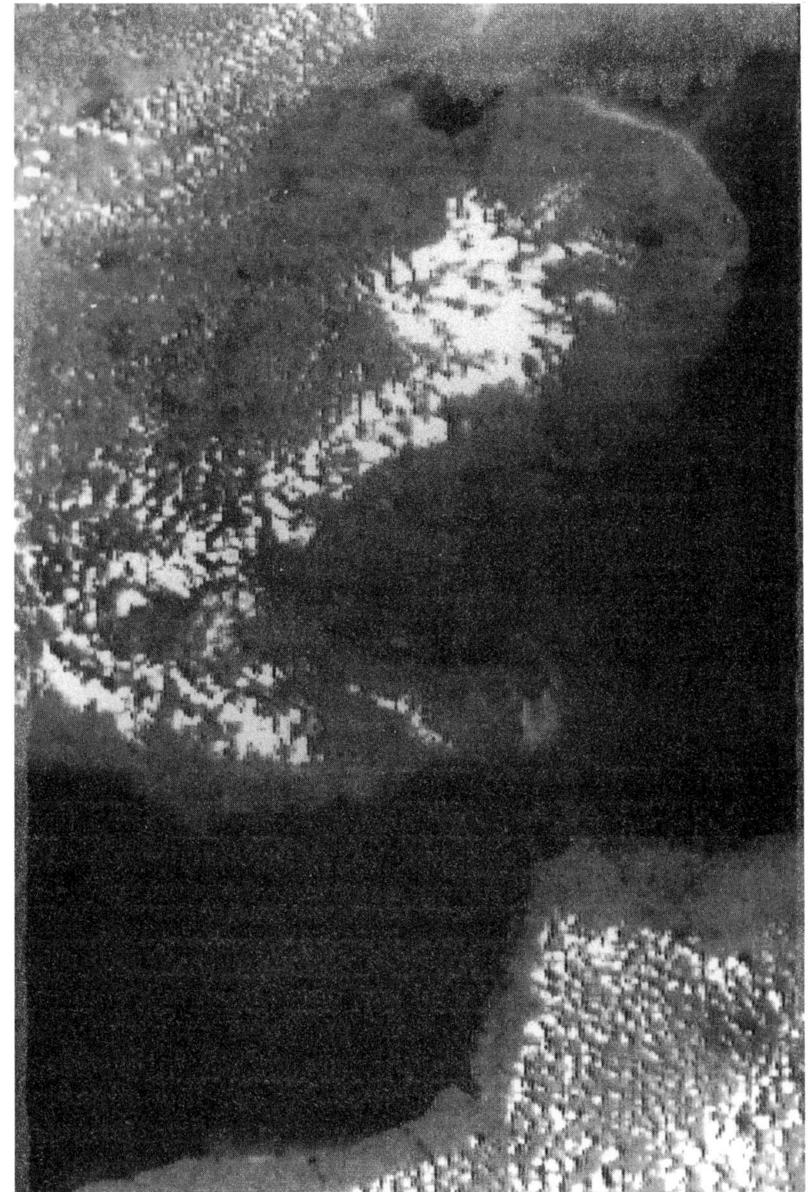

Figure 10.14. Cloud patterns show the sea-breeze which is blowing over southern England and the north coast of France. Coast fog can be seen along the coast to the top of the picture. (Courtesy of Dundee University.)

A satellite orbiting over the pole gradually covers the whole of the Earth's surface as the planet rotates beneath it during each orbit. Polar orbiting satellites in use include NOAA-9 and NOAA-10, which also transmit the three types of image. In the infra-red images the coldest (highest) cloud images are white, whilst the warmer (lower) tops are grey. In the visible images the densest cloud appears white and the thinner cloud, grey – This is probably the best type of image for showing sea-breeze clouds.

10.62 Routine satellite photography

Regular repeated photography of any particular part of the world now makes it possible to compare cloud forms near the coast on days of different sea-breeze situations. One such area, covering the north coast of France and part of south-east England where much observation material is available, is shown in figure 10.14.

The weather on this day, 8 July 1983, was sunny with very light easterly drift. When the photo was taken the sea breeze, shown by the strip 20 or 30 km wide completely free of cloud, was blowing from the English Channel towards the north coast of France and the south coast of England. The cumulus above France seem comparatively uniform, but over England there are several irregular features, most of which indicate well-known sea-breeze features.

An area of coastal fog can be seen in this picture along the Norfolk coast. Coastal sea fog, called haar in north-east Britain, is especially common over the cool waters of the North Sea, and frequently drifts inland on the sea breeze. In this case, however, as soon as the sea-breeze fog reaches the coast it is rapidly dispersed by heating over the land.

In the evening the infra-red pictures also show the effect of a sea breeze after the cloud has vanished. Where the sea air has penetrated, the land becomes distinctly cooler and the change of temperature shows up for several hours after sunset (Scorer, 1990).

11

Laboratory measurements

The meteorologist, in the study of complex flow patterns, is not often in the position of the physicist who constructs a controlled laboratory experiment. The former is a spectator of the weather, attempting to develop theories about systems whose initial conditions are imperfectly known and certainly uncontrollable. Flows which can be simulated in the laboratory with any degree of realism are therefore of considerable scientific interest.

As well as illustrating the general nature of flow patterns, laboratory experiments can give dimensionless measurements of velocities, densities and proportions of flows to be expected in the 'real world'.

11.1 'Land- and sea-breeze' simulation in water tanks

Experiments have been carried out in large water tanks in which a flow circulation results from the differential heating of the floor. The principle of these experiments is shown in figure 11.1.

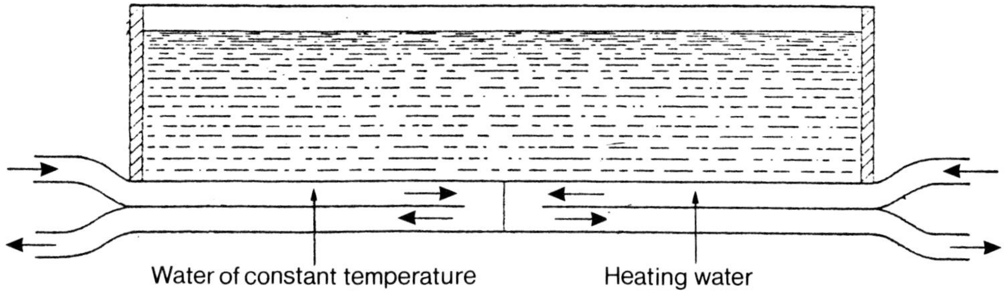

Figure 11.1. Sea-breeze simulation in a laboratory water tank. Part of the floor is heated and part is at constant temperature. (After Kobayasi, Sasaki & Osanai, 1937.)

One essential feature is a floor divided into two sections, which can be kept at controlled temperatures. The second feature needed is a stable temperature gradient maintained in the tank, for otherwise as soon as any circulation is started by heating it will extend throughout the whole depth of the liquid. Fine aluminium powder dusted into the water makes its motion visible and this motion can be photographed through the glass front of the tank.

Figure 11.2 shows two results from an early series of this type of experiment (Kobayasi, Sasaki & Osanai, 1937). This shows how the early streamlines formed ellipses whose centres shifted gradually towards the heated half. A circulation front, as shown in figure 11.2(*b*), formed 2 to 3 minutes after the heating was started and its velocity of propagation was found to be nearly uniform.

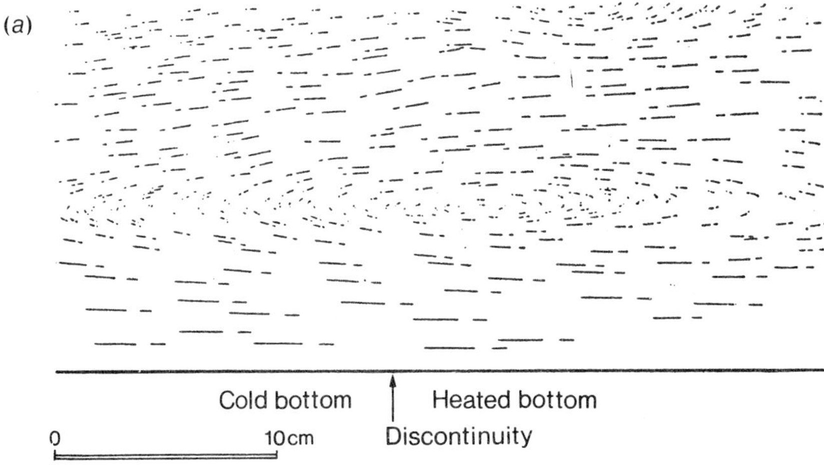

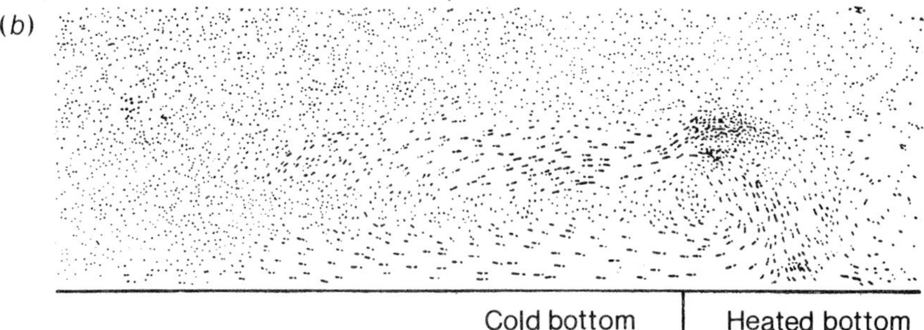

Figure 11.2. Streamlines in laboratory 'sea-breeze' experiments. The velocity of the fluid is shown by the length of the dash; the direction of movement is from the dash to the dot. (*a*) Early streamlines forming ellipses. (*b*) Front propagating at almost constant speed. (After Mitsumoto & Ueda, 1983.)

If cooling instead of heating was applied to the floor of the tank, the circulation in the reverse sense was much flatter than in the case of heating. The limits of the circulation on both halves of the bottom at any time were obscure, since the fluid began its motion very slowly.

11.11 Land and sea breeze with diurnal period

In a more recent series of experiments (Mitsumoto & Ueda, 1983) the complete diurnal period of land and sea wind was simulated. To do this the floor sections were given a regular variation to simulate several land and sea-breeze cycles.

The flow was visualised by a tellurium electrolite method. The black suspension emitted was a good tracer of the slowly moving flow as the settling speed of the particles was less than 0.1 mm s^{-1}.

When the land- and sea-surface temperatures are equal, the land breeze is still dominant. Soon after this, following the outbreak of convective cells over the land, which forms a thick mixed layer, the sea breeze intrudes over the land, with strong upward motion at its leading edge. Next, the intrusion of the sea breeze is deep over the land, and is accompanied by compensating reverse flow. Finally, the sea breeze weakens and the change to land breeze occurs.

Both the sea and land breezes were identified as gravity currents, but a clear difference was found between the profiles at their fronts. The land breeze advances against a headwind in stable stratified surroundings, so it has a flattened shape and the upcurrent at the leading edge is not as strong as at the sea breeze. On the other hand, the experiments suggested that the sea breeze advances into the mixed layer which has been generated by cellular convection and is accompanied by a strong updraft at the leading edge of the head. The advancing speeds of these fronts agreed well with observations of atmospheric outflows (Goff, 1976) and other laboratory experiments by Simpson & Britter (1980).

11.2 The generation of sea-breeze fronts

Two classes of laboratory experiment have been carried out which help our understanding of the formation of fronts (frontogenesis) in the sea breeze. The first type of experiment examines how horizontal temperature gradients work towards front formation and the second class sees how externally applied turbulence works against front formation.

All such experiments are best carried out in water tanks by releasing dense salt solutions into fresh water to produce the required density differences which are formed in the atmosphere by temperature differences.

11.21 Front formation from horizontal density gradients

It is helpful to consider whether a front can be formed by releasing a series of locks in each of which the density is successively reduced.

An experiment with at least three fluids is needed to produce a clear analogy of the continuous-gradient case. Figure 11.3 shows an experiment using three fluids of different densities. When barriers separating the fluids are removed a weak front forms at the leading edge of fluid 1. Fluid 2 begins to catch up this

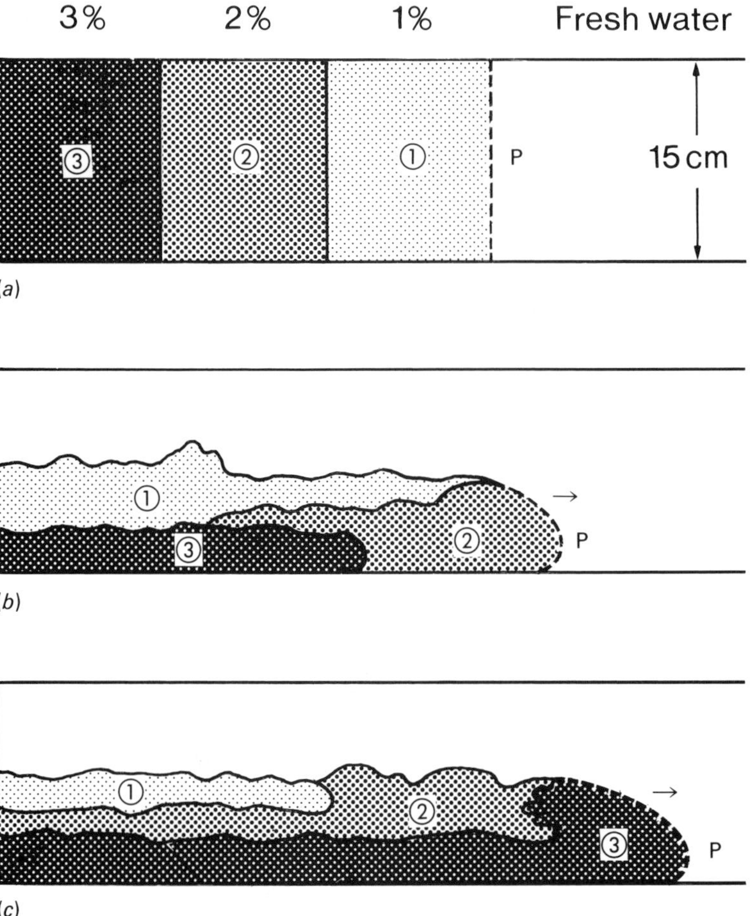

Figure 11.3. Flow after removing vertical barriers between three fluids labelled 1, 2 and 3, of different saline strengths 1%, 2% and 3% respectively. The density at the foremost boundary of the salt water, marked - - - -, increases from 1% in (*a*), to 2% in (*b*) and 3% in (*c*). P = foremost point of the front. A similar effect occurs when the initial variation of density is continuous, as in the sea-breeze front.

front and when it reaches it fluid 3 begins to catch up. Eventually it reaches the front, which then has a strength of 3%. We see that the front will only form if there is a change in the effective density gradient. This overall change occurs at the foremost point, *P*, where there is zero density gradient to the right and a mean gradient to the left of 1% per lock length.

11.22 Tilting-tank experiment for frontogenesis

Since it is difficult to set up a continuous horizontal density gradient in a fluid at rest, experiments have been performed in which a tank is filled while in the vertical position with a vertical density gradient and then rapidly rotated through 90 ° to the horizontal position (Simpson & Linden, 1989). Figure 11.4 shows

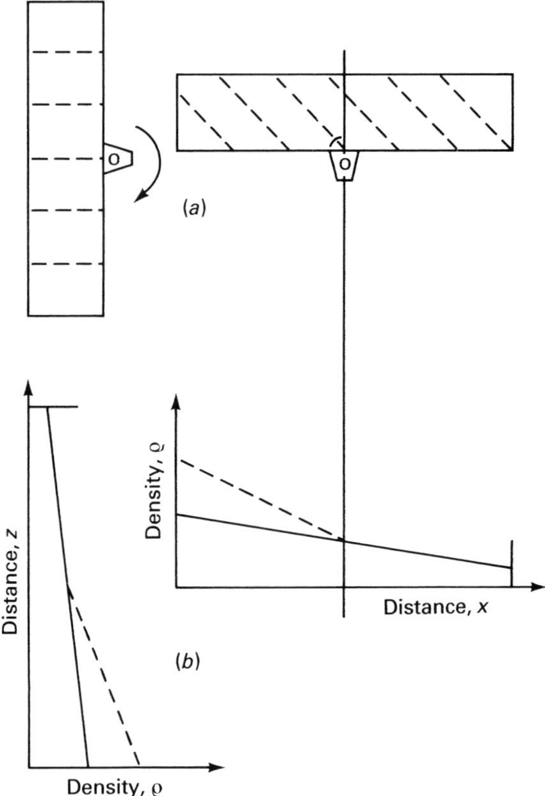

Figure 11.4. (*a*) The tilting-tank experiment. The dashed lines are lines of constant density before and after tilting the tank. (*b*) Density profiles. The solid lines show density profiles from the tilting experiment in (*a*). The dashed line shows density profiles from another experiment, showing different gradients in the two halves. These are capable of generating a front.

the principle of these experiments, in which a closed rectangular perspex tank was used, 3.6 m long, 150 mm deep and 80 mm wide. In Figure 11.4(*a*) the lines of equal density are seen after a rapid rotation through 90 °: the fluid will only move through an angle of about 60 °. In figure 11.4(*b*) the principle of one experiment is shown in which the density profile was greater in the left half, but there was no actual change in density at the centre point. Thus we have two different horizontal gradients exactly as in the atmospheric observations shown in figure 3.5.

The photographs in figure 11.5 show three stages in the development of a front in a tilting-tank experiment in which the density contours are indicated by dye lines. The initial stage of frontogenesis is seen in the curvature of dye lines, a hint of which can be seen in the top photograph. When the change in density gradients is large enough a strong front is formed, as shown in the photographs. These photographs are taken by a shadowgraph method in which parallel light is shone from a projector through the flows and seen on a translucent screen in front of the tank. The optical effects formed by the variation of refractive index with density of the fluid form a very powerful way of showing local density gradients.

11.23 Turbulent mixing and frontogenesis

The formation and destruction of a gravity current front in the presence of turbulence have been examined in laboratory experiments by Linden & Simpson (1986). Figure 11.6 shows the apparatus, a rectangular perspex tank two metres long, with a removable vertical partition across the centre separating two fluids of different density. The fluids are kept turbulent by bubbling air from the base of the tank.

When the barrier is removed the dyed dense fluid begins to flow underneath the light fluid and a gravity current front is initially formed, as shown in figure 11.7(*a*). As the gravity current moves along the tank the turbulence mixes it with its surroundings; this process can be seen in figure 11.7(*b*). Eventually, the fluid becomes completely mixed vertically and the density gradients are horizontal, as shown in figure 11.7(*c*).

In some of these experiments the turbulence was turned off while there was still an appreciable horizontal density gradient, and under these circumstances the re-formation of a sharp gravity current front was observed. Measurements made in this experiment with a conductivity probe showed that the dense fluid just behind the front had originated from the far end of the tank and had thus travelled the full length of the tank with little vertical mixing. It appeared that the non-uniformity of the horizontal density gradient after the turbulent mixing had ceased was responsible for this frontogenesis.

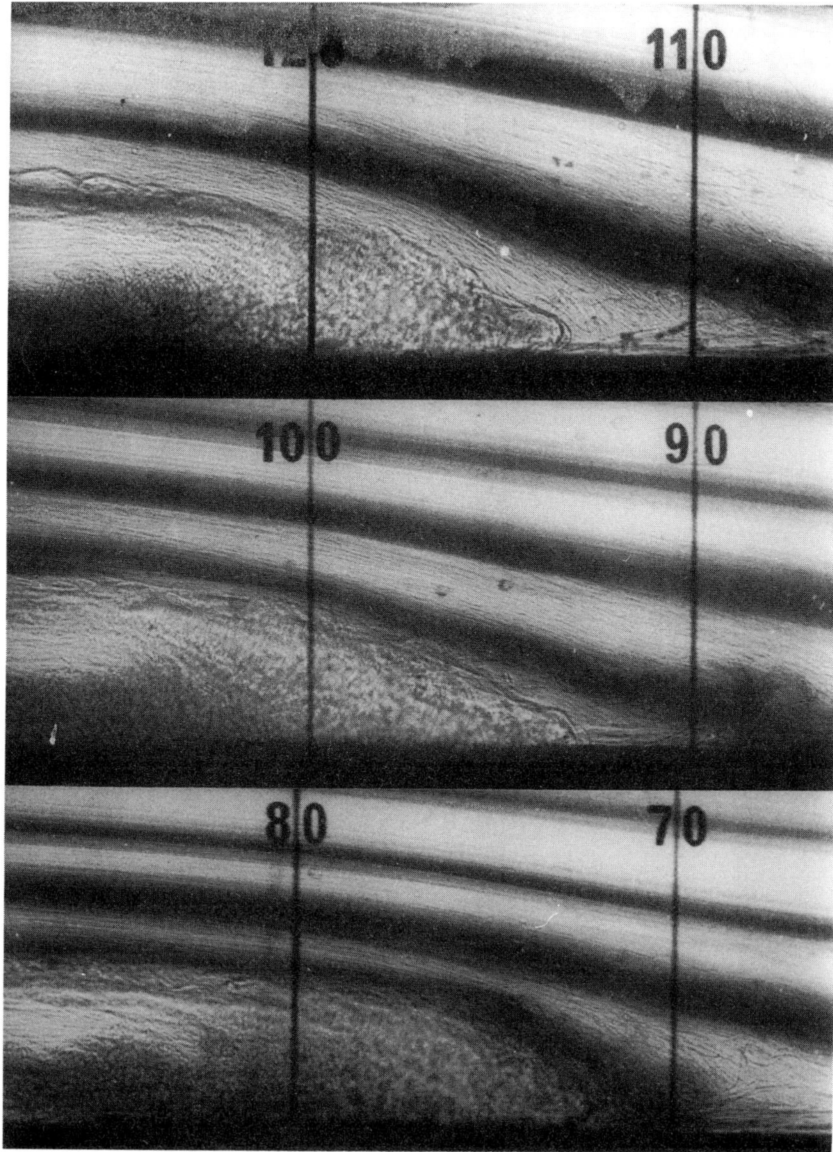

Figure 11.5. Three stages during frontogenesis in the experiment shown in figure 11.4, 9, 11 and 13 seconds after the tank reached the horizontal. The flow is from left to right.

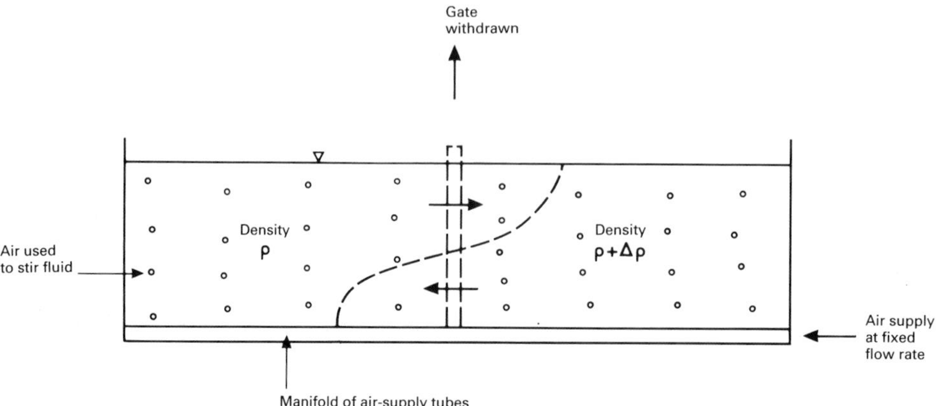

Figure 11.6. Apparatus to show the formation and destruction of a gravity current front in the presence of turbulence. Air bubbles rising from the floor are used to stir the fluid.

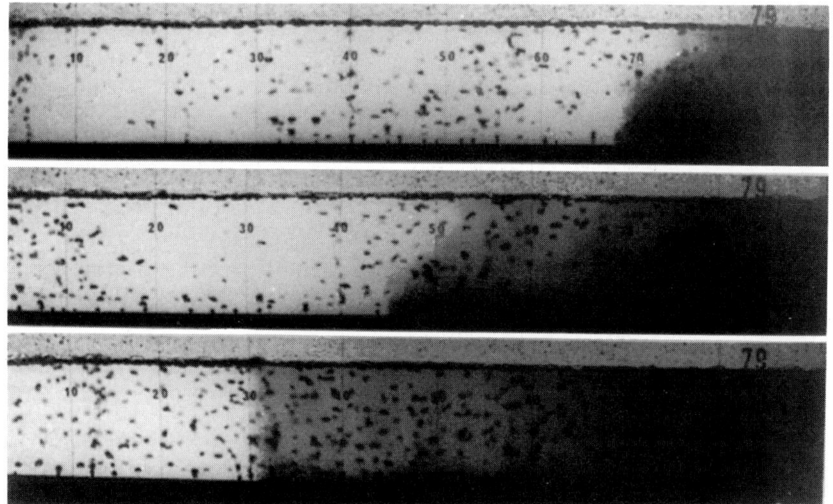

Figure 11.7. A dense flow moving from the right through turbulence which mixes it with its surroundings.

11.3 Sea-breeze fronts

11.31 Gravity-current experiments in water tanks

Gravity currents are formed whenever a dense fluid advances through surroundings of lower density. Experimental work in this field has a long history; gravity

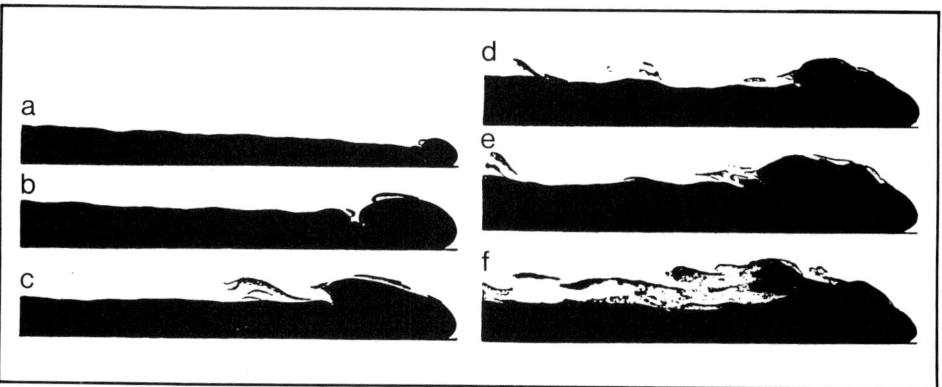

Figure 11.8. Shadow pictures showing profiles of the head of a gravity current of cold water in a hot water tank. The temperature difference is very small in (*a*) and increases to 35 °C by (*f*). (After Schmidt, 1911.)

currents have been studied in lock exchanges between salt and fresh water and are particularly important in the spread of dense gas from accidental releases.

Gravity-current fronts appear in the atmosphere as thunderstorm outflows and as sea-breeze fronts. An attempt to model these flows, using water tanks, was made as long ago as 1911, when Schmidt (1911) used cold water flows through surrounding hot water in small laboratory tanks. Figure 11.8 shows his shadow pictures of the profiles of the head of a gravity current, with density differences increasing from views (*a*) to (*f*).

In experiments on fluid flow the value of the dimensionless Reynolds number has been shown to be important, since its value gives a measure of the effect of viscosity on the nature of the flow. For flows at velocity U and depth h, it can be shown that the Reynolds number

$$\mathrm{Re} = Uh/\nu$$

where ν is the kinematic viscosity, is non-dimensional. The dimensions of U are L/T; h is L, and ν has dimensions L^2/T, the same as Uh, so Uh/ν has no dimensions.

The Reynolds numbers of the flows in Schmidt's photographs can be calculated and are found to range from about 5 in figure 11.8(*a*) up to 1000 in (*f*). As the Reynolds number increases the shape of the head of the gravity current alters, the foremost point of the nose approaches the ground and more intense mixing occurs at the front and top of the head. The final view, figure 11.8(*f*), shows a profile which is found to be typical of all flows with Re greater than 1000, which have a turbulent wake streaming back behind the head.

Experiments have also measured the dimensionless value of the velocity of a gravity-current front as a function of Reynolds number (Keulegan, 1957). In

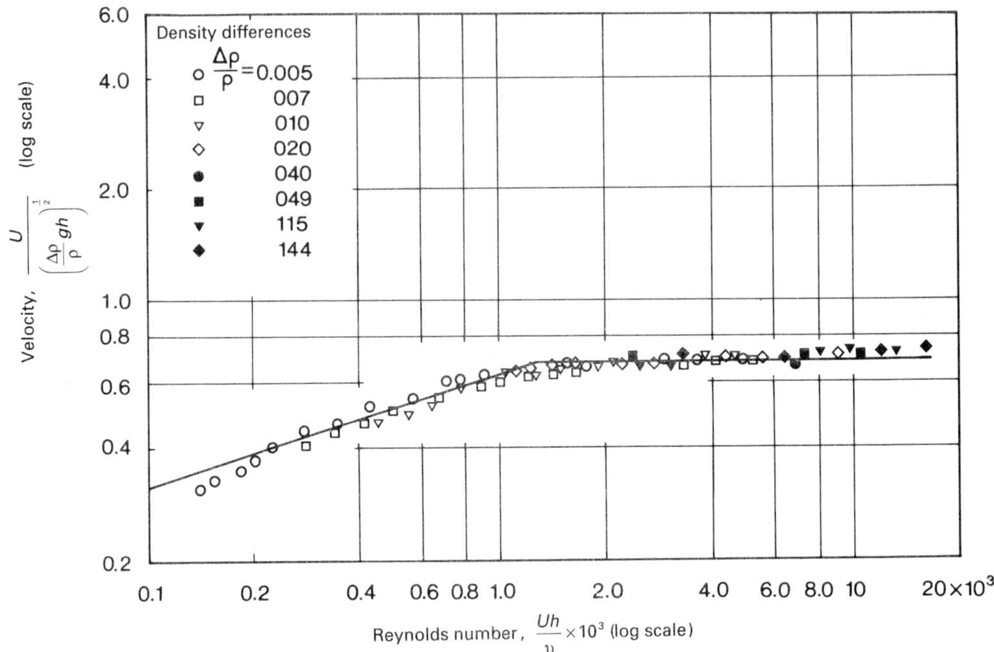

Figure 11.9. The dimensionless velocity of a gravity-current head for different Reynolds numbers. Both axes are in logarithmic scales. This velocity changes very little for Re >1000. (After Keulegan, 1957.)

figure 11.9 the velocity is written in dimensionless form as $U/(\delta\varrho/\varrho\, gH)^{\frac{1}{2}}$, and is plotted against Re. It can be seen that the dimensionless velocity does not depend on Reynolds number for Re values greater than 1000.

11.32 Appearance of the front

The appearance of an advancing gravity-current front in a water tank is shown in figure 11.10. That sea-breeze fronts have the same form can be seen when they are carrying sufficient pollution to make the whole profile visible.

The complex shifting forms seen at the front can be analysed into two main forms of instability, the billows on the one hand and lobes and clefts on the other.

Figure 11.10. The form of a gravity current of salt water advancing in a fresh-water tank. The density difference is 1% the height of the flow is about 10 cm and its form is made visible by a milky dye. Lobes and clefts are clearly visible.

11.33 Billows

The billows form an important mechanism in the mixing at the front. An unusual view is given in figure 11.11, a laboratory photograph of dyed ambient fluid above and beneath a dense gravity-current head. Kelvin–Helmholtz billows form above and are made visible by shining a narrow slit of light into fluorescein mixed with the ambient fluid.

11.34 Lobes and clefts

The general view of the front in figure 11.10 shows advancing lobes which are divided by deep clefts. As the front advances with its overhanging nose, some of the less-dense ambient fluid must be overrun. This leads to a highly unstable

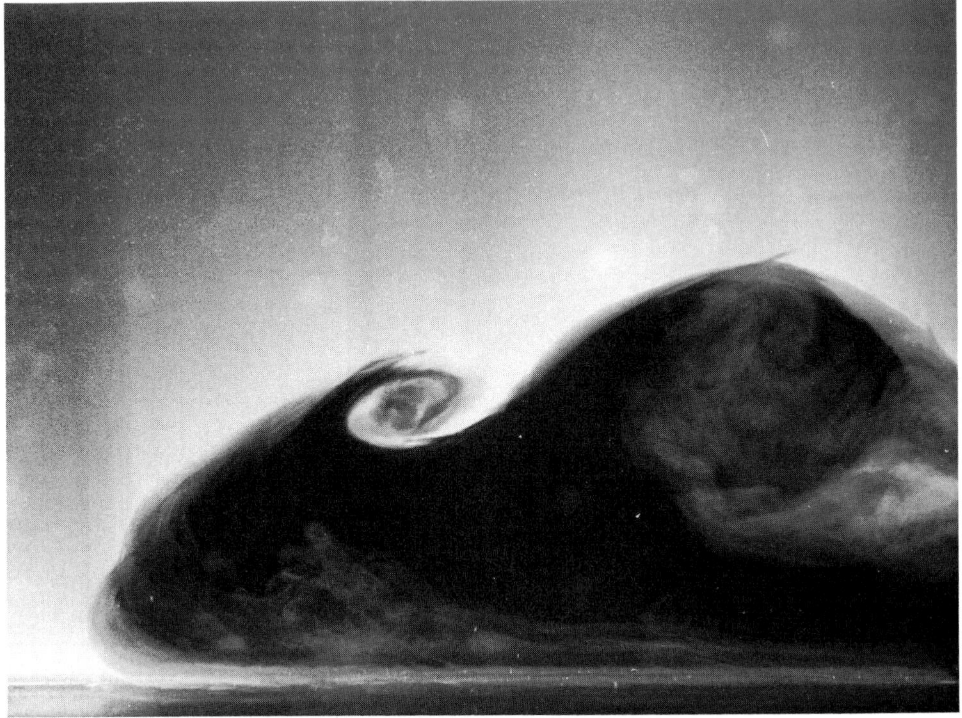

Figure 11.11. Laboratory photograph of less-dense fluid mixing above and below a gravity-current front. Kelvin–Helmholtz billows form above; fluid ingested from below rises as it approaches the leading edge.

state near the front, as this overrun light fluid makes its way upwards and towards the leading edge of the front. This effect can be seen in figure 11.11, where some dyed fluid can be seen rising from the base of the tank.

Successive shadowgraphs in plan view at intervals of about $\frac{1}{2}$ second are able to show the evolution of lobes and clefts. Figure 11.12 shows such a series of views. Some lobes expand to a maximum width when they divide and a new cleft is formed. Other lobes contract and vanish as two adjacent clefts meet. As a result, the number of lobes remains roughly constant, since they all reach about the same maximum breakdown size; in the sequence shown the number of lobes remains constant at about seven.

Measurements of the maximum lobe size, given as a fraction of the head height, show a decrease as the Reynolds number increases up to a value of 1000 (Simpson, 1972). For values of Re greater than this a steady value is found for the maximum lobe size of about half of the total head height; the mean size is about half this. Although very few field measurements exist for lobe size at sea-breeze fronts, numerous photographs of dust-laden thunderstorm cold out-

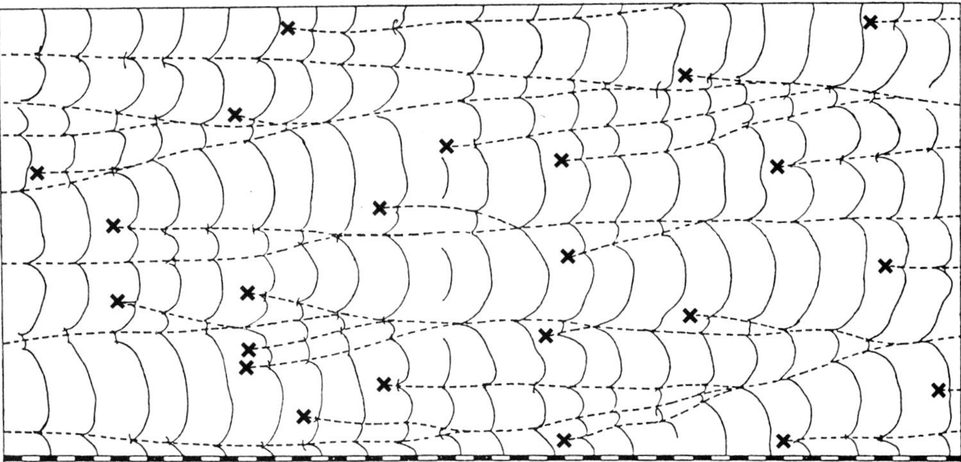

Figure 11.12. The evolution of lobes and clefts at a gravity-current front seen from above. X marks the points where new clefts appear, dashed lines show its subsequent progress. The scale is in centimetres, the total water depth is 24 cm.

flows suggest the mean size of $\frac{1}{4}$ to be a reasonable value in these large Reynolds number flows.

11.4 Use of tanks with moving floor

Some simple early experiments used a moving floor in the tank to modify the effect of the lower boundary on a moving gravity current (Simpson, 1972). It was found possible, by appropriate motion of the floor, to reduce the height of the foremost point or nose and hence the amount of overrunning of less-dense fluid at the head. This inhibited the formation of lobes and clefts and formed a truly two-dimensional flow.

More ambitious experiments were carried out by moving the floor of the tank and at the same time altering the speed of the ambient fluid. Figure 11.13 shows a schematic diagram of this apparatus (Britter & Simpson, 1978; Simpson & Britter, 1979). The object of the experiments was to bring the gravity current to rest in the laboratory, so that the head could be studied as near as possible in a steady state. Water was pumped through a transparent working section one metre long, having a conveyer-belt floor which could be moved in the same direction as the water flow. A metered saline flow was introduced in the opposite direction from beneath the floor. Water and mixed salt solution discharged over a down-stream weir of adjustable height. By suitable adjustment of flow and

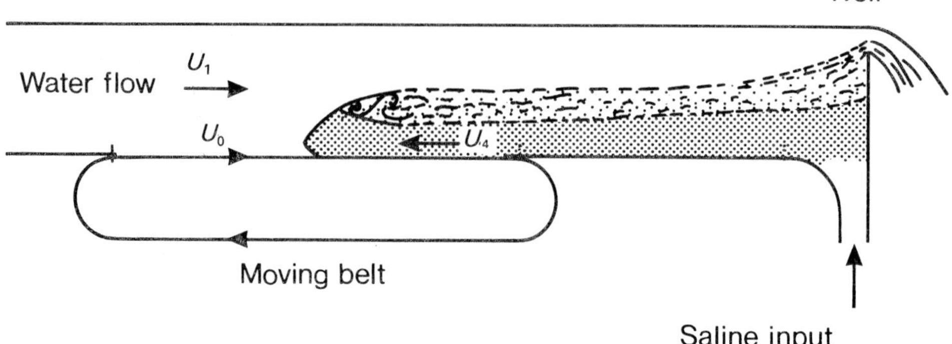

Figure 11.13. Tank with moving floor used to investigate the dynamics of a gravity-current head. The current is brought to rest above the moving floor and can be controlled by the opposing flow and the dense fluid input (shaded). U_o = floor speed, U_1 = opposing flow and U_4 = flow in the gravity current, (shaded).

floor speed the saline flow could be brought to rest, usually half way along the moving belt.

11.41 Head and tail winds

To model the effect of ambient winds on the behaviour of the gravity-current head the opposing water flow in the tank was run at a different speed from that of the floor. When the flow was faster than that of the floor the effect was that of an opposing, or head wind. Figure 11.14 compares the head profiles of steady-state gravity-current heads brought to rest in conditions of head-wind, calm and tail-wind. It can be seen in the shadowgraph photographs that for a head wind the profile is flatter, and the height of the nose is lower, in proportion to the head height than in calm surroundings. The speeds of advance of given gravity currents into known head winds were measured and compared with a set of atmospheric measurements (Simpson & Britter, 1980). It was found that the gravity-current advance was reduced by 0.6 times the opposing flow. The best agreement was obtained for gravity currents of about a quarter of the total depth of the ambient fluid.

11.5 Ambient stratification: two-layer system

The formation of an undular bore by a sea-breeze gravity current was described in section 3.5. It was shown that the dense layer of air which had formed near

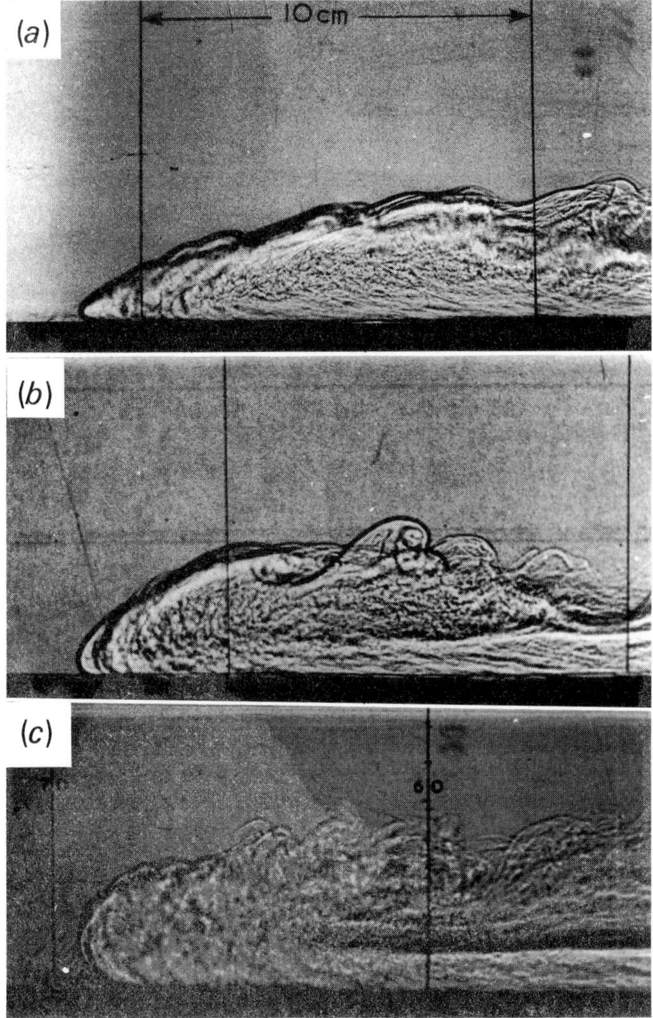

Figure 11.14. Profiles of gravity-current heads brought to rest in an experimental tank. (*a*) Head wind (*b*) calm (*c*) tail wind.

the ground in the evening completely changed the entire nature of the front as it began to change into the advancing waves of an internal bore. Similar phenomena occur as thunderstorm outflows move through stable layers of air (Fulton, Zrnic & Doviak, 1990; Koch *et al.*, 1991).

One of the simplest types of ambient stratification studied in the laboratory is that of two layers, in which one fluid of constant density lies above another of greater density, with a sharp interface between them.

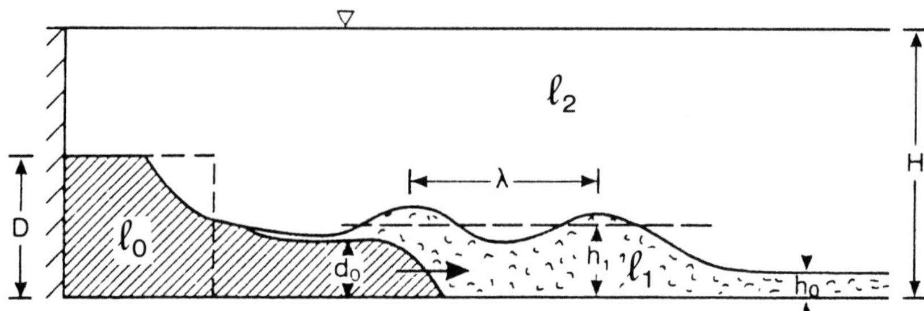

Figure 11.15. A gravity current of density ϱ_0 forming an obstacle in a two-layer system. An undular bore is moving ahead of the obstacle.

Experiments have measured the results of moving a solid obstacle through a two-layer system (Wood & Simpson, 1984) and more recently such work has extended to the effect of a gravity current considered as a more complex obstacle (Rottman & Simpson, 1989), using apparatus shown in diagrammatic form in figure 11.15. The two-layer system consists of fluid of density ϱ_2 lying above fluid of greater density ϱ_1. A gravity current of greater density ϱ_0 is released into this system and effects are observed of flows for different speeds and depths.

The different possibilities and the results obtained from experiments are plotted in the theoretical diagram in figure 11.16, which shows the types of disturbance produced by smooth obstacles of different speed and size. Both the speed and the size of the gravity current obstacle are conveniently plotted in non-dimensional form – the velocity, F_0 as a fraction $U/(g'h_0)^{\frac{1}{2}}$, the Froude number, and size D_0 as a fraction of d_0 and h_0, where h_0 is the depth of the lower dense layer, d_0 is the depth of the obstacle and $g' = (\Delta\varrho/\varrho)g$.

11.51 Subcritical and supercritical speeds

The points to the left of the graph represent the effect of obstacles whose depth is small compared with the depth of the lower fluid. The effect produced by these smaller obstacles consists only of a simple displacement either upwards or downwards, according to whether the flow is supercritical or subcritical, that is, according to whether its velocity, U, is greater or less than $(g'h_0)^{\frac{1}{2}}$, which represents the maximum speed of small internal waves on the interface. Another way of stating this is 'whether the internal Froude number $U/(g'h_0)^{\frac{1}{2}}$ is greater or less than 1'.

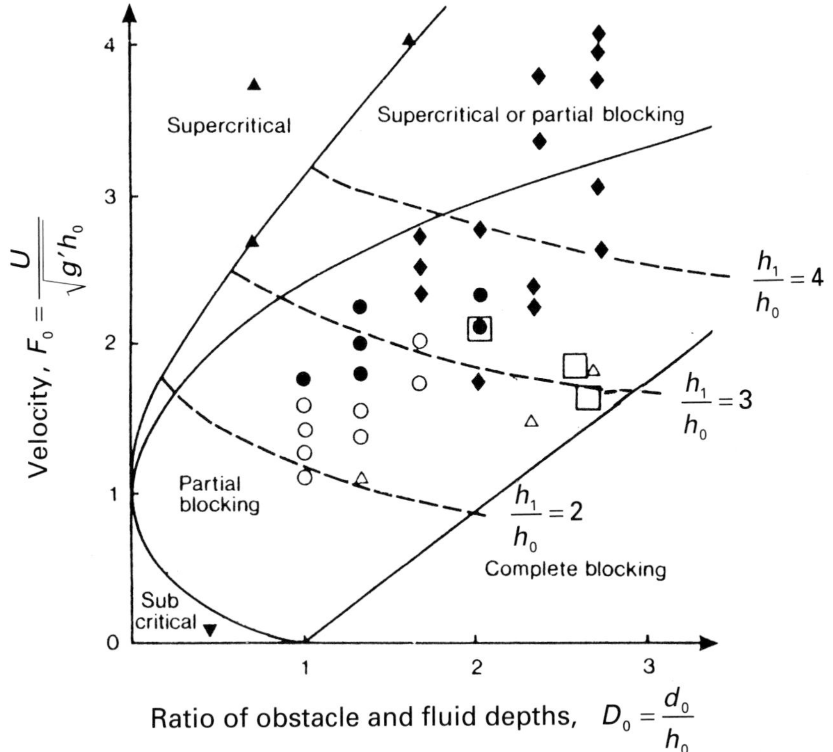

Figure 11.16. Velocity (given as Froude number F_0) of different types of flow observed when an obstacle of height d_0 is moved in a fluid of depth h_0. The type of flow are ▼ subcritical gravity current. ▲ supercritical gravity current. ○ ● undular bores. ◆ turbulent bores. △ intrusion. □ atmospheric measurements.

11.52 Flows with partial blocking

Blocking may be taken to mean the movement of the flow upstream at the speed of the obstacle. Complete blocking is not possible from an advancing gravity current, but partial blocking can exist in which the flow over the obstacle is reduced as some of the fluid is forced to move ahead of it in the form of a bore. A sketch of this process is shown in figure 11.17.

The first stage in the generation of an undular bore is the formation of a smooth hump which envelops the head of the gravity current. This hump moves forward along the interface between the two layers, leaving behind it some of the denser fluid of the gravity current. The gravity current is disrupted, but

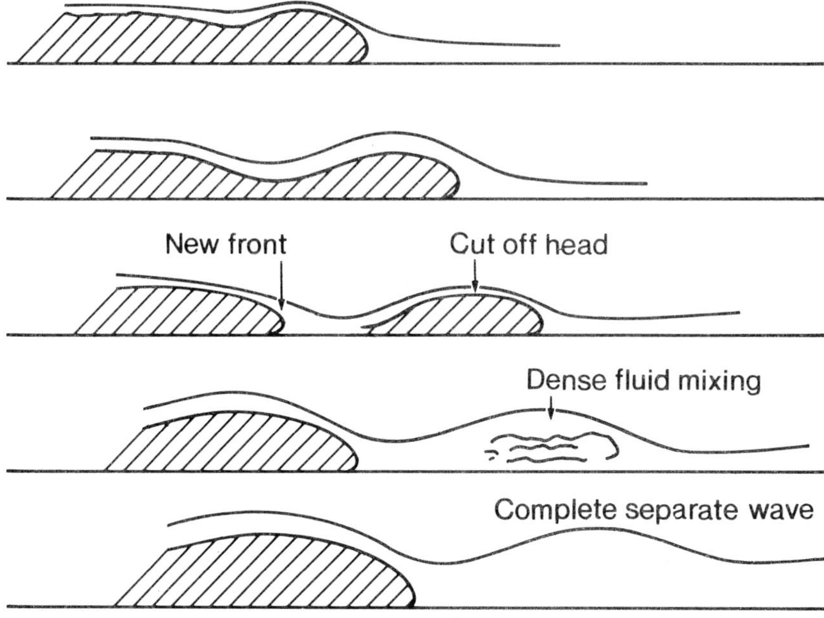

Figure 11.17. Stages in the generation of an undular bore by a gravity-current front moving though a dense layer at the bottom of a water tank.

a fresh head soon forms, a fresh hump forms above this, and the process is repeated.

The results of a series of such experiments also appeared in figure 11.16. The bores are of undular form for values of the bore-strength h_1/h_0 up to about 3.5, and for greater bore strength it takes the form of a turbulent bore, looking much like a gravity-current front. Three atmospheric examples of undular bores are also plotted on the same graph.

11.6 Collisions of fronts

Several cases have been described in which two sea-breeze fronts have collided. When the collision has been at a fairly small angle it results in a convergence zone with much increased convection.

Collisions in which the two fronts have approached each other almost head-on are especially interesting, because in some cases the two sea breezes have appeared to cross and then proceed in their original directions as two separate fronts, at first sight just as if they were unaffected by the meeting.

The head-on collision has been modelled as a two-dimensional flow in a rectangular laboratory tank by releasing simultaneously a gravity current from a lock at each end of the tank.

11.61 Collision of two identical fronts

When two identical fronts collide, the flow is symmetrical about the centre plane on which they meet so the flow pattern is the same on each side as when a single front collides with a wall.

These simple wall collisions are easy to carry out in experiments. When the gravity current reaches the wall it climbs up it to a certain height and then comes instantaneously to rest. The heap of dense fluid then falls down into the two-layer system which it had established in its previous motion. These conditions form a solitary wave or an internal undular bore, which can be seen moving steadily away from the wall. The rate of advance of this disturbance is close to that of the original gravity current before the collision.

11.62 Collision of two unequal fronts

The result when two unequal gravity currents collide is similar; the main effect is the emergence of two bore waves in opposite directions. Only a few experimental results have been published (Kot & Simpson, 1987), but a numerical study by Clarke (1984) has displayed this result very clearly.

Figure 11.18, shadowgraphs of two colliding fronts, shows the domed shapes of the two different fluids which are initiating the bores in the two opposite directions. The progress of two colliding fluids in another experiment is displayed in figure 11.19. The different parts of the collision process are outlined with the speed and positions of the fronts and bores.

The approaching gravity currents are about the same depth, but the front on the left contains the denser fluid and its velocity U_1 is greater than the velocity U_2 of the other front. After the collision a small amount of the fluid from the front at the left continues to move along the lower boundary. The energy has almost all been transferred to bore B_2, which moves at about the same speed as U_2. Looking to the left-hand side, the main flow is the bore B_1, with a small intrusion F_2 above it. The more energetic bore B_1 moves at a higher speed, similar to the original speed U_1.

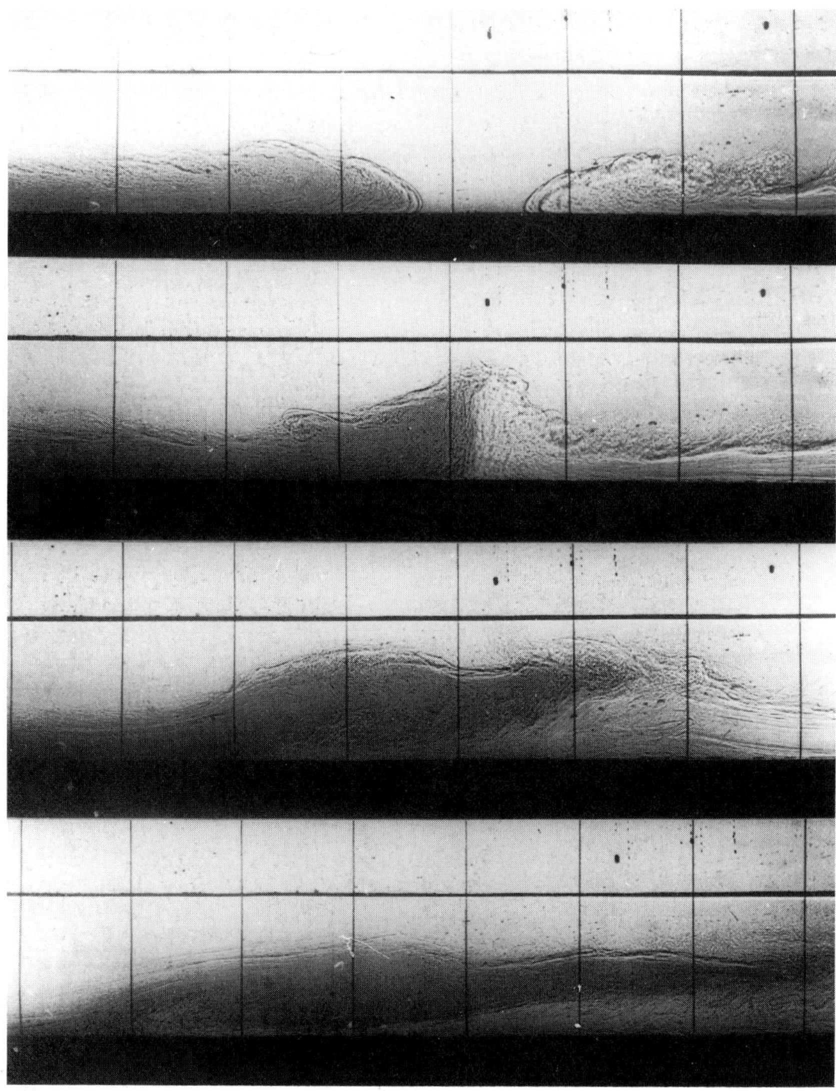

Figure 11.18. Shadowgraphs of two colliding fronts.

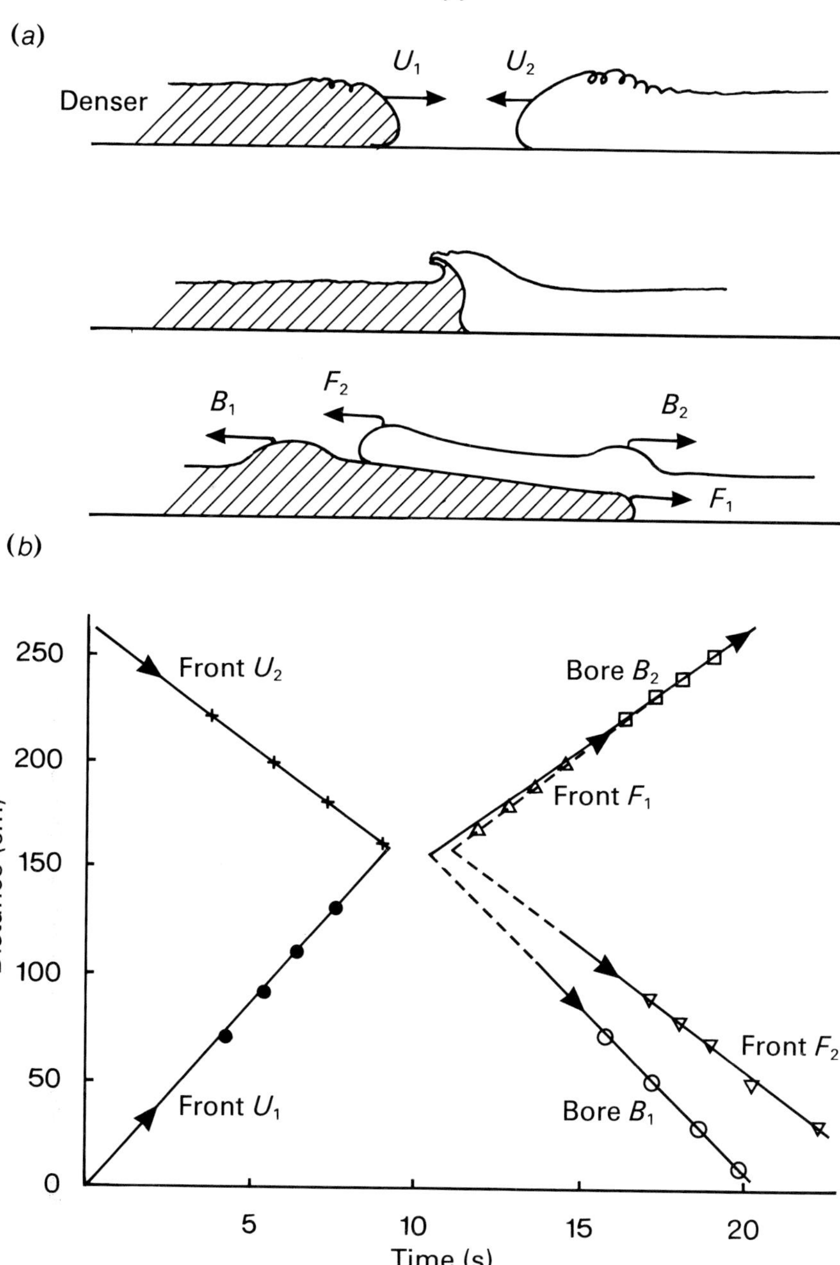

Figure 11.19. (*a*) Two colliding fronts with velocities V_1 and V_2 resulting in fronts with velocities F_1 and F_2 and bores with velocities B_1 and B_2. The positions of the approaching fronts are plotted to the left. On the right are seen the two new fronts and the bores formed at the collision. (From an unpublished experiment by Higson & Simpson.)

11.7 Requirements of laboratory models

The requirements for formal dynamic similarity include exact geometric similarity between the model (laboratory gravity current) and the real world (sea-breeze front).

The geometric profile of a model gravity-current head was illustrated in figure 11.9, in which it was shown to change with increasing Reynolds number until a steady profile is reached which is independent of Reynolds number. The requirement here is that the Reynolds number should be greater than a fixed number (of about 1000) for viscosity effects in the model to be insignificant and geometric similarity to be obtained in the flows.

This dimensionless form of a gravity-current head in the conditions of calm surroundings established in the laboratory models has been found to agree with measurements of sea-breeze fronts in the field (Britter & Simpson, 1978; Simpson & Britter, 1979).

Dynamical similarity can be established by showing that the Froude number (a measure of the ratio of inertial forces to buoyancy forces) of an observed sea-breeze front was in general agreement with those measured for laboratory gravity currents. The Froude number is also found to be independent of the Reynolds number, provided it is greater than 1000 (figure 11.9).

Laboratory experiments are unable to examine the whole range of complex stratifications present in the atmosphere, but, as we have seen, the simplified versions of stratification used in many experiments have succeeded in giving useful information.

At the time of writing there are many gaps still to be filled in the study of over-running less-dense fluid, the effects of velocity profiles near the ground and the nature of the turbulent mixing at the front. These gaps exist not only in the laboratory experiments but also in the extent of the field work carried out.

11.71 Interpretation of laboratory results

We have seen above how the laboratory can be used to illustrate the general nature of flow patterns which can occur under controlled conditions, but also to give measurements of velocities, densities, and proportions of flows to be expected in the 'real world'.

A final example of a simple series of experiments and their results is given in figure 11.20. This examines the height, h_n, of the foremost point, or nose, of the profile of a gravity-current head. The measurements of this height have been

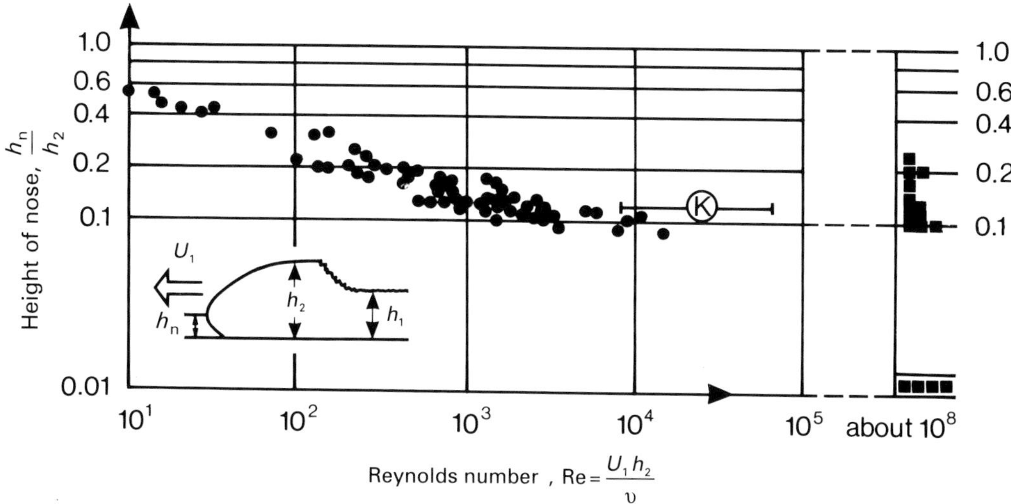

Figure 11.20. The dimensionless height of the nose (or foremost point) of gravity-current fronts for different Reynolds numbers, in the laboratory and in the atmosphere. The fractional height is almost constant for all Re >1000. Laboratory, ● Simpson, K Keulegan. Atmosphere ■.

obtained in about 60 experiments in the laboratory in which the Reynolds numbers, U_1h_2/ν, vary from 10 to just above 10^5.

The results show that the nose height, given as the dimensionless number h_n/h_2, lies between 0.1 and 0.2, provided that the Reynolds number is greater than 10^3. On the right of the graph are some results measured from sea-breeze fronts and thunderstorm outflows, which show a good agreement with the experimental results.

11.72 Conclusions

Many excellent laboratory results are now available, and more could certainly be usefully carried out. However, they have difficulty in representing all the properties required and a more complete description will come from a combination of observations, laboratory experiments and numerical modelling.

12

Theoretical models

Both analytic and numerical models of the sea breeze have been developed.

In analytic models the object is to understand the essential physics of the sea-breeze circulation, such as what determines its velocity and its horizontal extent. This can be done by introducing reasonable simplifications to the governing equations and parameterising various physical processes. These simplifications in fact give a significant insight into the dynamics of the sea-breeze circulation. However, it is these simplifications which limit the applicability of analytical models to the real atmosphere. To model the sea breeze realistically in the atmosphere, one has to consider various complex processes, heat balance at the ground, heat transfer from the ground, topography, and so on. This only became possible after the electronic computer was reasonably well developed.

Numerical models try to solve a set of complicated equations which govern the motion of the atmosphere, the transfer of heat and the heat balance at the Earth's surface. Since computer capacity is finite, one has to calculate the flow field at discrete grid points, though the equations have the information for the entire flow field. Numerical modellers are still faced with the problem of adequately resolving the steep temperature and velocity gradients which occur near the coast, while nevertheless maintaining lateral boundaries far away to avoid contaminating the solution.

Numerical models give a realistic simulation of the actual sea-breeze circulation; however, they may not help to obtain a physical understanding of the phenomena.

The factors which affect the land and sea breeze circulation are:

1. Diurnal variation of the ground temperature
2. Diffusion of heat
3. Static stability
4. Coriolis forces

5. Diffusion of momentum
6. Prevailing wind

The first three factors are essential. The fourth, fifth and sixth factors are not necessary in producing the sea breeze. However, the fourth factor plays an important part in determining its horizontal dimension and producing the clockwise rotation with time. The fifth factor plays an important part in producing a realistic wind profile near the ground. The sixth factor plays a significant role. If it is very strong, the sea breeze cannot be generated. If it is moderate the sea breeze front is known to be formed.

12.1 Analytic models

Most of the analytical models depend on so-called linear theory. The linear theory assumes that the amplitude of the diurnal variation of the ground temperature is much smaller than the vertical temperature difference between the ground and the height affected by the sea-breeze circulation, and neglects several terms in the governing equations.

The first to treat the problem hydrodynamically was Jeffreys (1922) who considered the land and sea breeze to be forms of antitriptic wind. This means that the only forces taken into account are the two due to friction and to the pressure gradient caused by the unequal heating.

The values of the variation of velocity with height from Jeffreys' paper are shown in figure 12.1. The strength calculated was of the order of magnitude of an actual sea breeze and showed the decrease with height which is a feature of sea breezes. It shows that above a certain height the wind blows towards the warmer side. The height of zero wind was 150 metres, but depended on a number of variable factors such as eddy viscosity and the lapse rate. The theory seemed to give a reasonable first approximation to the observed velocity profiles, but, since Jeffreys neglected the effects of time variations in the governing equations, his theory gave the daily variation of wind in phase with the daily variation curve of temperature. This does not occure in nature.

Because the sea-breeze circulation is generally restricted to the lowest 1–2 km of the atmosphere, it is strongly influenced by the boundary layer processes of viscosity and conduction. Its time scale is too long for the Earth's rotation to be ignored, except near the equatorial region. Solutions in a number of analytic models have been obtained only after simplifying the governing equations to the point where certain of the physical processes listed above were omitted.

Heat conduction was omitted by several workers, who specified the heating as a function of space and time. Notable examples are Haurwitz (1947) and Schmidt (1947), who both specified the pressure gradient normal to the coast

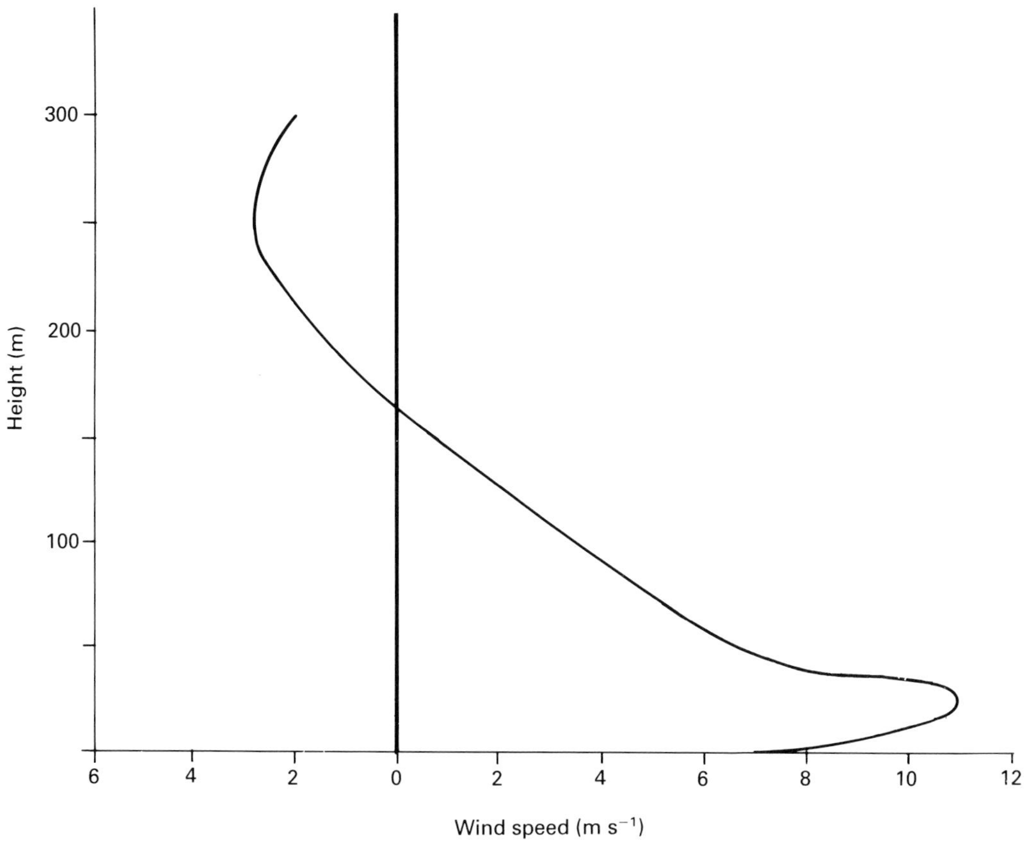

Figure 12.1. Variation of velocity of the sea-breeze with height. (After Jeffreys, 1922.)

as a sinusoidal function of time. Haurwitz specified the pressure gradient as independent of both horizontal distance, x, and vertical distance, z, and by including both Coriolis and frictional forces was able to obtain development of realistic wind hodographs with time. Figure 12.2 shows two hodographs obtained by Haurwitz for different values of the coefficient of friction k. For a number of reasons the hodographs actually observed are not nearly so regular as those shown in the figure, but they have the same general form. Schmidt prescribed the pressure gradient as a sinusoidal function of x and time, t, which exponentially decays with height.

Even in Schmidt's approach nothing can be said about the spatial scale of motion except that it must be prescribed as a function of x, z and t.

Defant (1951) specified the surface temperature as a sinusoidal function of time and space. He included the diffusion of heat from the ground, but the

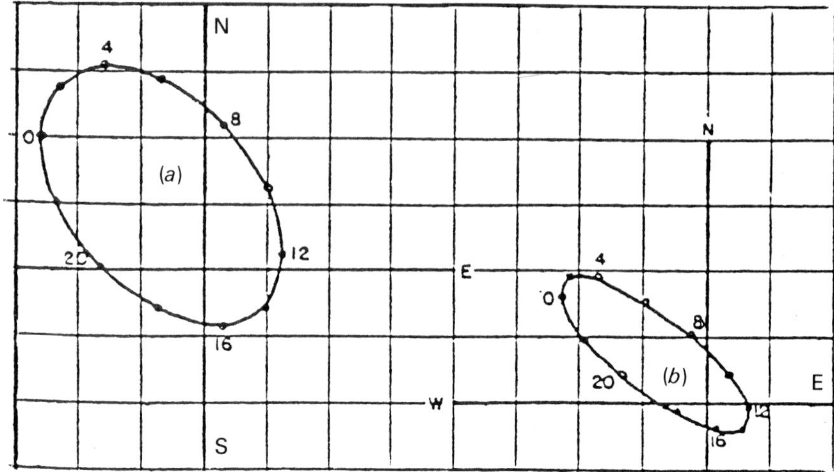

Figure 12.2. Effect of friction, k, on the sea-breeze hodograph. (a) $k = 0.58 \times 10^4\ s^{-1}$. ($b$) $k = 10^4\ s^{-1}$. The side of a square corresponds to $1\ m^{-1}$. (From Haurwitz, *J. Meteorol.*, 1947.)

diffusion of momentum was neglected for simplicity. He was the first to solve for the flow in the x–z plane incorporating Coriolis force.

12.12 Incorporation of all six physical factors

A linear theory was presented by Walsh (1974) which included all the six factors listed above in section 11.1. The motion was forced by a prescribed surface temperature function, and a sea breeze was produced with realistic velocities and spatial dimensions.

The dependence of the circulation on Coriolis forces was examined for the case of no prevailing wind; the winds rotated in the expected clockwise sense at all values of z. At large x and z the dependence on the Coriolis parameter, f, was very noticeable. Alterations of the stability showed that, near the coastline, the low-level wind was strongest when the stability was weakest, suggesting that the observed weakness of the land breeze could be attributed to the higher night-time stability of the atmosphere.

Results were obtained including the basic gradient wind, U, which has an important effect in advecting the perturbation quantities. Values of the speed of this wind were imposed which varied from 1 to 10 m s^{-1}. A series of results appears in figure 12.3, which shows the x-component of velocity at $z = 0.33$ at three different times, for the case when the offshore gradient wind is

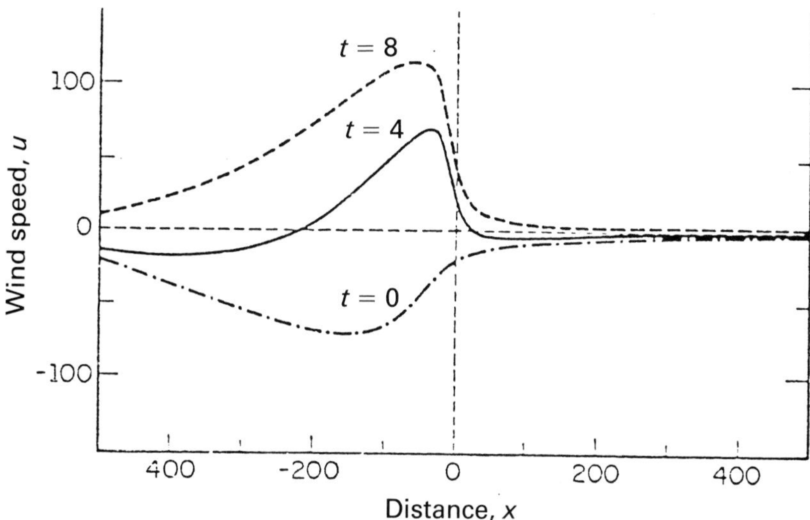

Figure 12.3. Strength of the x-component of the sea-breeze at $z = 0.33$ at three different times, when the offshore gradient wind is 5 m^{-1}. (From Walsh, *J. Atmos. Sci.*, 1974.)

5 m s^{-1}. This wind pattern is suggestive of the sea-breeze front which is observed to develop with an offshore wind. Walsh suggested that, although it has been reproduced in some non-linear numerical models, this asymmetry of winds may be the closest reproduction of a sea-breeze front that the model will allow.

This model was applied to the prediction of the development of the sea breeze with an opposing wind. As described above in Chapter 4, empirical relationships have been used for forecasting the onset of sea breeze which depend on the temperature difference and the strength of the opposing wind. Walsh's model provided theoretical support for the observed linear relationship between U^2 and the temperature difference (ΔT) max (vertical land–sea temperature difference). His diagram appears in figure 12.4; the agreement of the criterion lines is very close.

The relative intensities of the sea and land breeze mentioned above were further investigated by Mak & Walsh (1976). Their model included simple diurnal variations of both stability and of eddy diffusivities, giving the results illustrated in figure 12.5. These are very similar to those observed during the day in the absence of any synoptic wind. Diffusivity variations were considered to give less effect than stability fluctuations.

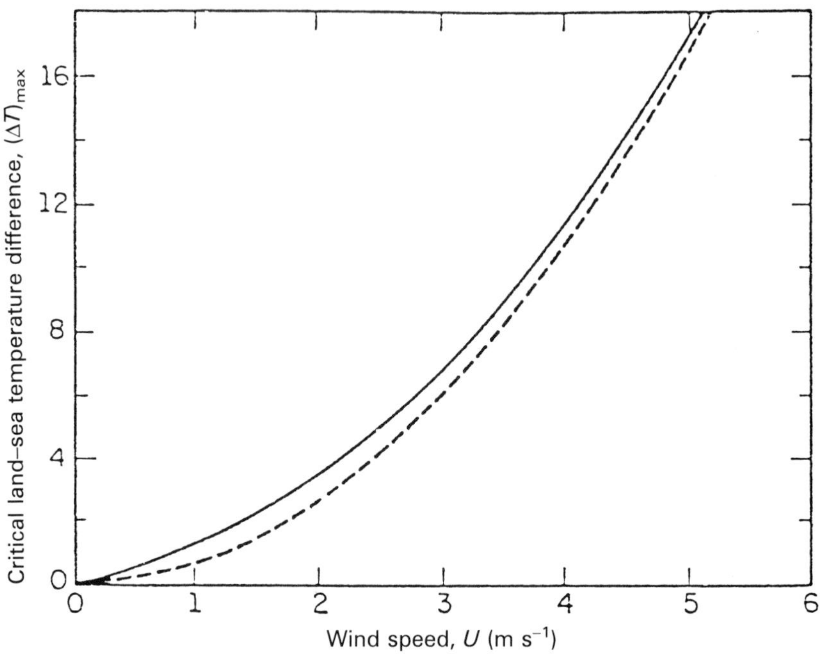

Figure 12.4. The critical land–sea temperature difference required for sea-breeze occurrence, (solid curve). The dashed curve is based on the observational data of Biggs & Graves (1962). (From Walsh, 1974).

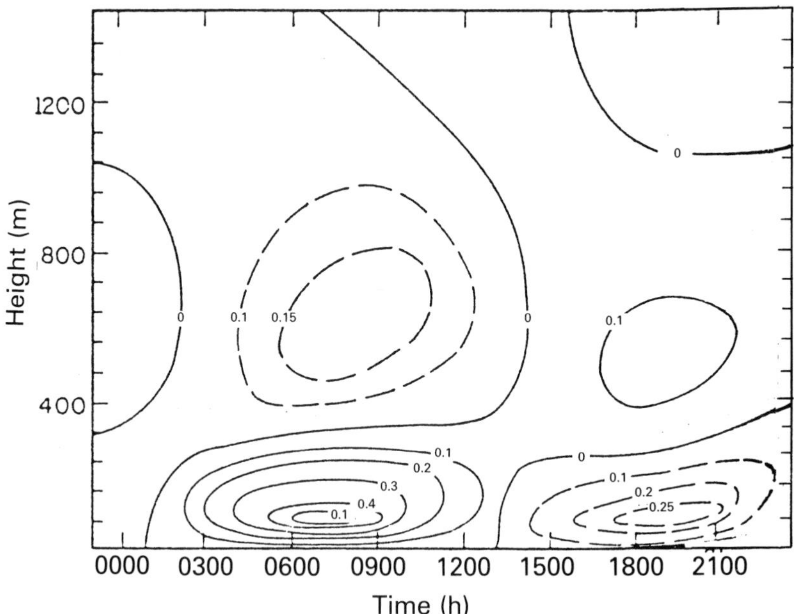

Figure 12.5. Model including diurnal variations of both stability and of eddy diffusivities, to show the difference between sea breezes (left) and land breezes (right). The contour lines show wind speed. Onshore breeze is shown by solid lines and offshore wind by dashed lines. (From Mak & Walsh, *J. Atmos. Sci.*, 1976.)

12.13 Horizontal scale of the motion

What determines the horizontal scale of the land- and sea-breeze circulation has received only scant attention, but a key to this question was given by Kimura & Eguchi (1978), who studied the sea-breeze development over an island of various widths. They found that when the width of the island is small the horizontal scale of the sea breezes which develop at both sides of the island is comparable with the width of the island. The sea breeze starts without any phase lag behind the temperature rise, as in the work of Jeffreys (1922). When the width of the island is large the horizontal scale is much smaller than the width of the island implying that there is a horizontal scale of sea breezes which is internally determined.

Rotunno (1983) tried to investigate this using an inviscid model, but because of this neglect of viscosity he obtained infinite horizontal scale at the inertial latitude where the diurnal frequency coincides with the Coriolis parameter (which happens at the latitude of 30 °). This does not happen in reality.

If viscosity is suitably introduced it is possible to overcome this defect. Niino (1987) used the same model as Walsh (1974) but introduced a different scaling under which the flow fields for various values of external parameters collapse to the same single pattern. The intensity of the sea breeze decays with a finite distance from the coastline and the distance is scaled by (N/w) $(k/w)^{\frac{1}{2}}$ where N is the buoyancy frequency, K the thermal diffusivity and w the diurnal frequency. This means that the horizontal extent of the sea breeze increases with increasing atmospheric stability and thermal diffusivity. Figure 12.6., which is based on the work of Niino, shows the dependence of the horizontal scale with latitude for typical atmospheric values of N and K. Due to the variation of Coriolis force with latitude the spread of the sea breeze decreases from 70 km at the equator to about 25 km at high latitudes.

12.14 Limitations of linear theory

Linear theory does not include the change with time of the vertical stratification of the atmosphere, nor the changes in diffusivity with height. One of the consequences is that the sea-breeze and land-breeze circulations are symmetric and that the formation of the sea-breeze front cannot be modelled. To overcome these and other limitations, numerical models have been developed.

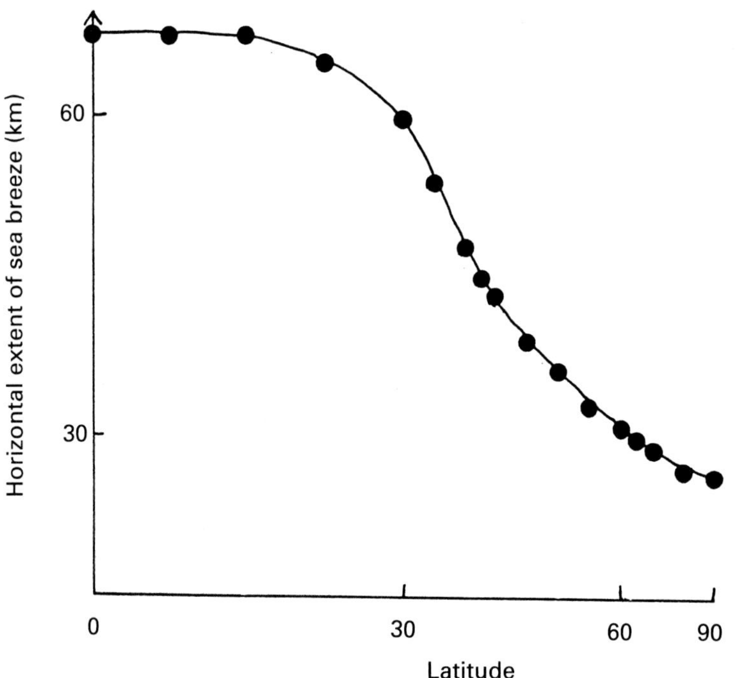

Figure 12.6. Horizontal extent of sea breeze plotted against latitude, showing the effect of the Coriolis force. (After Niino, 1987.)

12.2 Numerical models

As described above, the linearisation of the equations of motion was necessary to ease their solutions. The primary omission was of the advective terms, removing any feedback effect between the velocity and temperature distributions.

With the advent of the computer the calculation of solutions to the non-linear equations by numerical methods become possible. An early attempt was made by Pearce (1955). In contrast with the works described above, which started with assumed temperature distributions, Pearce postulated a distribution of the heating and cooling sources and allowed an initially isothermal atmosphere to be heated by convection currents, establishing an adiabatic lapse rate throughout a certain height.

Estoque (1961) employed a different approach, and his ideas were used in several models which followed. The vertical cross-section of his area was bounded at the top by the level $z = 2$ km, and the lateral boundaries were 100 times greater. He divided the entire region into two horizontal sublayers: a thin stratum 50 m deep where the vertical fluxes of heat and momentum were con-

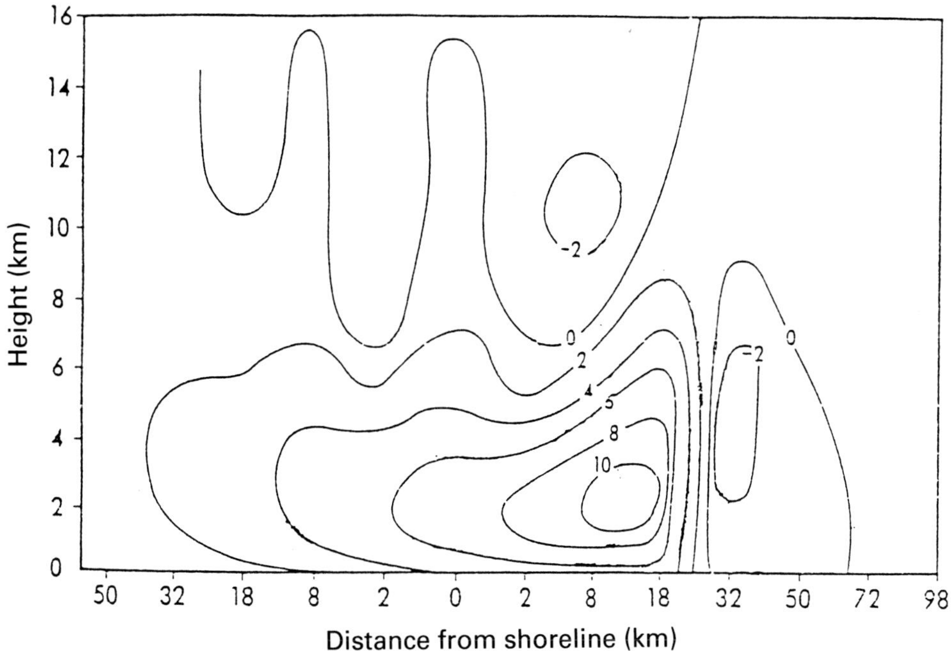

Figure 12.7. Vertical cross-section of a sea breeze after eight hours simulation at 1600 h, showing contours of wind strength in m s^{-1}. (After Estoque, 1961.)

stant with height, and an overlying transition layer where the effect of eddy fluxes decreased with height. A vertical cross-section of the wind field in his model after 8 hours simulation is shown in figure 12.7 The model showed a low-level convergence zone between 18 and 32 km from the shore at this time, probably the nearest the model could approach to the sea-breeze front, the first time this had appeared in sea-breeze simulations. Another interesting feature in the model is an increase in low-level wind late in the evening, separated from the main sea-breeze. This feature was unknown in 1961, but has since then been shown to occur in the atmosphere.

The development of sea-breeze circulations under various types of prevailing large-scale wind conditions and thermal stratification was incorporated into this model by Estoque (1962). The model made clear the great difference between the influence of onshore and offshore winds, and agreed in many respects with observations.

The model was extended and applied to three dimensions by McPherson (1970), who also changed the profile for the mixing coefficient *K* in the 'transition zone' from a linear to an exponential one. He simulated the effects of a square bay, roughly equal to Galveston Bay, Texas, and found interesting

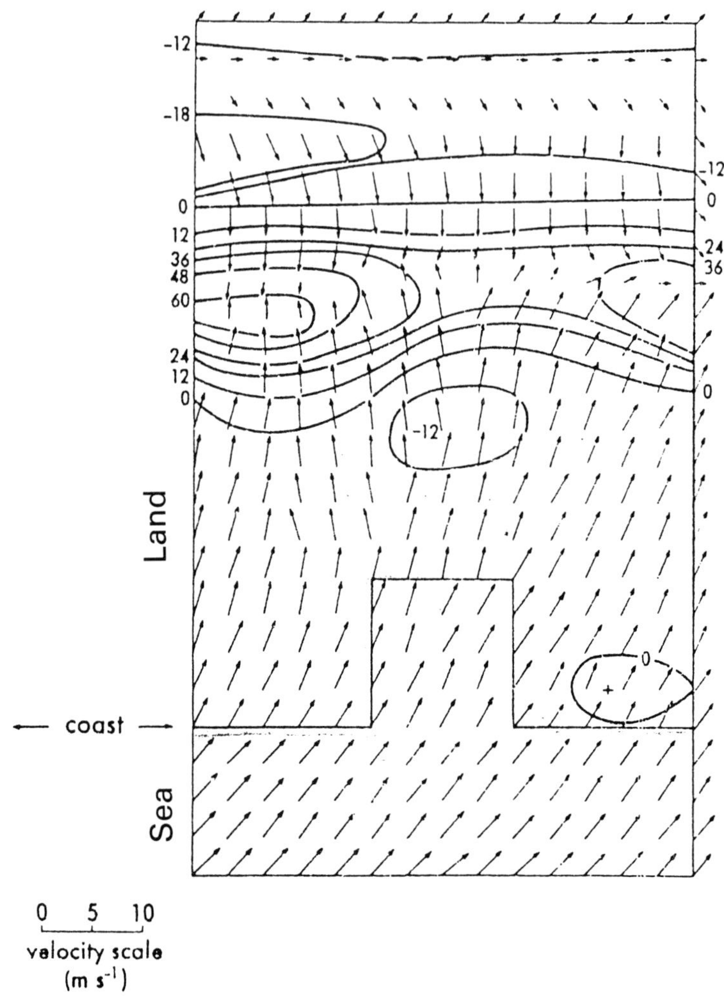

Figure 12.8 Theoretical flow over a regular shaped bay, after 10 h simulated time. The arrows give the magnitude (in m s^{-1}) of the flow at a height of 250 m. The solid lines show the vertical motion (in cm s^{-1}) at 850 m. (From McPherson, *J. Appl. Meteorol.*, 1970.)

deformations in the three-dimensional flow. As shown in figure 12.8, considerable asymmetry developed in the later stages of the sea-breeze. McPherson suggested that the convergence zone generated to the west was due to the fact that the Coriolis force and overall pressure force acted in the same direction on one side of the bay, but in opposite directions on the other side.

Further modifications allowed Neumann & Mahrer (1971, 1974) to investigate the diurnal variation. They were able to examine the formation of the land breeze and also the breeze round circular islands and circular lakes. They suggested that their horizontal grid spacing of 5 km might have been too large to specify the sea-breeze front in any detail. This seems to have been confirmed by the findings of Lambert (1974).

Lambert (1974) used the model of Estoque, but with a grid of horizontal spacing of 1 km, vertical spacing of 75 m and a time step of 40 s. He simulated the sea-breeze front in calm conditions and with an offshore geostrophic wind of 3 m s^{-1}. The results for the rate of advance of the front in these different conditions are of especial interest, as they have features in common with the observations available. Figure 12.9 shows his results for the advance of the front, replotted for comparison with the field observations of sea-breeze advance in southern England by Simpson, Mansfield & Milford (1977) shown previously in Figure 3.11.

Important steps in the application of a numerical model of the sea breeze to realistic coastlines were taken by Peilke (1974), who simulated the sea breeze over Florida. He used an eight-level, three-dimensional model, whose results

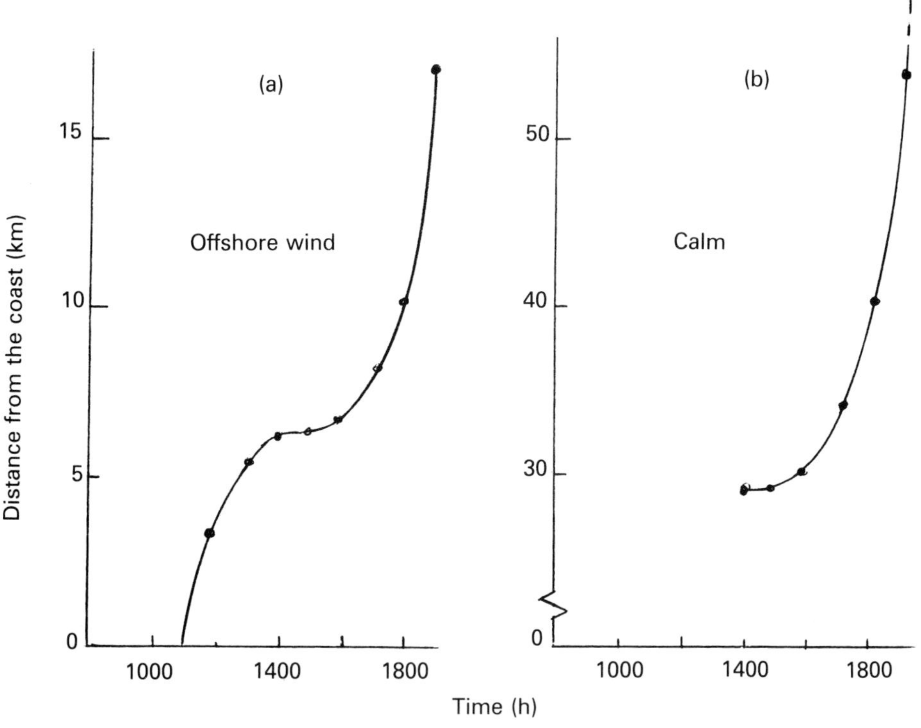

Figure 12.9. Hourly positions of the sea-breeze front from the model of Lambert (1974). (*a*) with offshore wind (*b*) in calm conditions. The results have features in common with observations shown in figure 3.11.

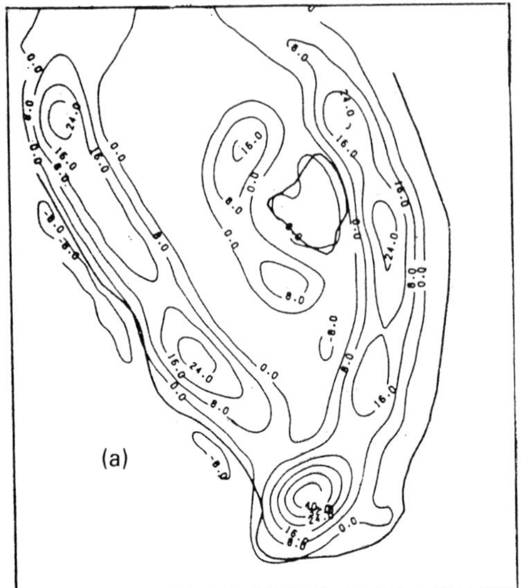

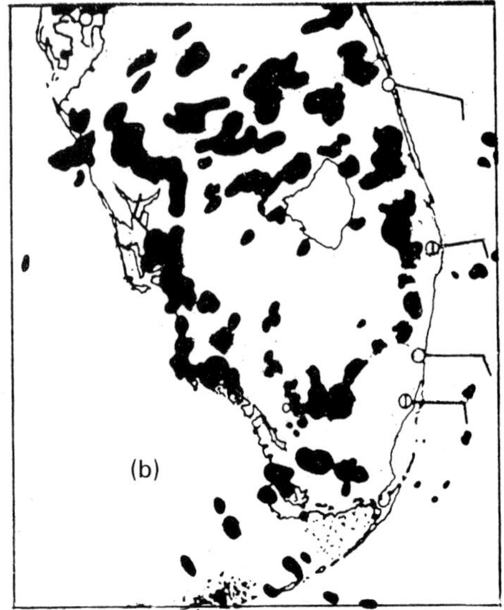

Figure 12.10. The sea breeze in Florida on 29 June 1971.
(*a*) Calculated vertical velocity, in cm s^{-1}, 7.5 h after sunrise. Geostrophic wind is 2.5 m s^{-1} from 110°.
(*b*) Showers seen from radar at Miami. 7 h 28 min. after sunrise.
A close parallel can be seen between the patterns of forecast rising air and the observed showers. (From Pielke, *J. Atmos. Sci.*, 1974.)

showed a close correspondence with both radar and satellite pictures of cloud development over Florida. An example is shown in figure 12.10, with a radar picture of shower clouds for comparison. The close correspondence between these two patterns of rising air led Pielke to conclude that, on days without significance synoptic disturbance, the sea-breeze convergence patterns were the primary control of the cloud and shower complexes in southern Florida.

The effects of topography to modify the sea-breeze circulation were included in a model by Mahrer & Pielke (1977). Their results showed that, depending on the separation of the mountains from the coast, the combined sea-breeze and mountain circulations can be more intense during both day and night when they act together.

12.21 More recent complex models

An experimental forecast using a mesoscale model covering the whole of England and Wales was produced by Carpenter (1979). This used a 10 km grid, imposed orography and the synoptic situation for 14 June 1975. The develop-

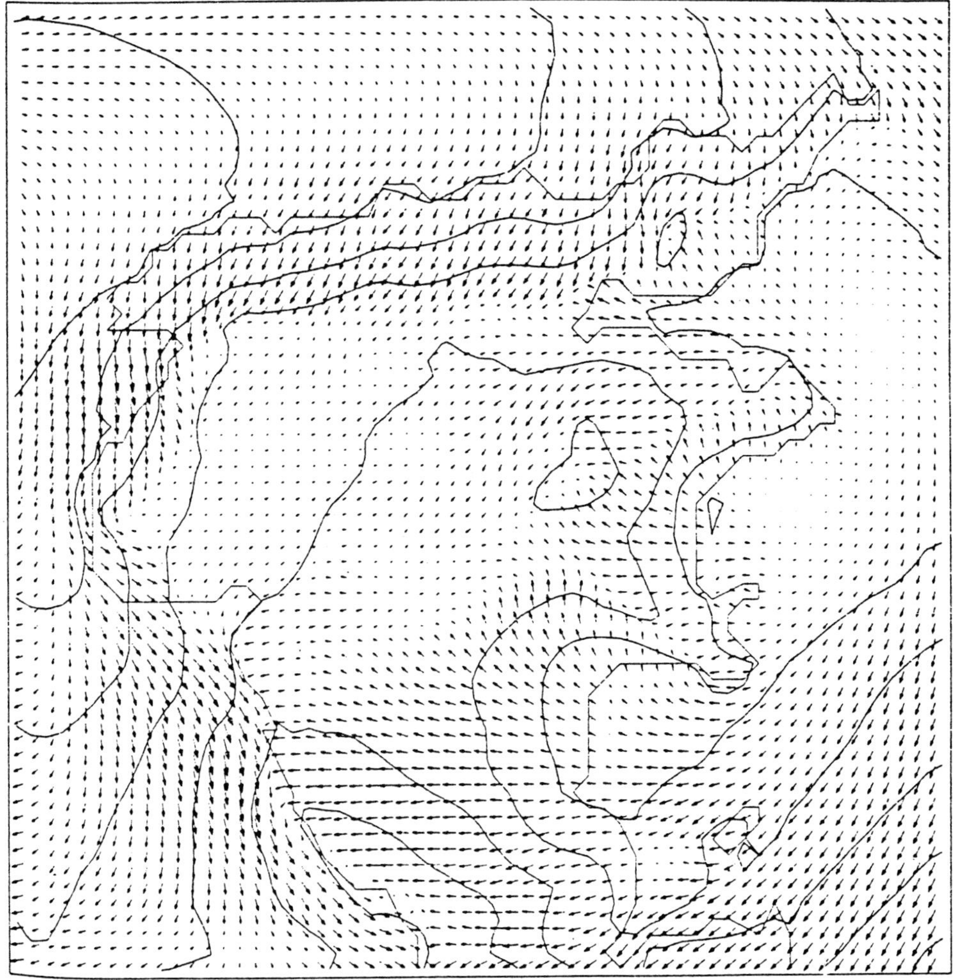

Figure 12.11. The 50 m wind and sea level pressure forecast for 1800 GMT on 14 June 1973 when both orographic effects and the movelent of the synoptic-scale antocyclone are included. The isopleth interval is 0.5 mb. (From Carpenter, *Q.J.R. Meteorol. Soc.*, 1979.)

ment of the pressure and wind fields closely followed the observations of Simpson, Mansfield & Milford (1977) made on that day. Figure 12.11 shows one of their forecast charts, for 1800 GMT, showing the wind at a height of 50 m and sea-level pressure; a close resemblance can be seen to the actual observations, which have been described in Chapter 3.

Physik (1980) used an 8 km grid and endeavoured to model correctly the relative importance of the latent heat and sensible heat (Bowen ratio) during the advance of the sea-breeze front. He also attempted to model the sea breeze in southern England on 14 June 1975. The model simulated well the average speed during the daytime and also the evening penetration.

For the future, it seems that even finer grid resolution will help to resolve further some of the questions still left unanswered by numerical models.

Some recent numerical studies of the development of the sea-breeze front by Garratt & Physick (1986,1987) are of special interest. They simulated gravity current flows in the atmosphere at mesoscale (20–200 km). In contrast to some of the earlier laboratory experiments which dealt with gravity currents in a steady state, they examined the effects of turbulent heat transfer from the ground and also that of the Earth's rotation; studying the rates of change of horizontal gradients normal to the front of temperature and velocity. The wind field was strongly dependent on the pre-frontal winds and on the magnitude of the pressure jump, and it was found that a vortex of dimensions of less than 50 km formed in the foremost part of the flow.

References

Aldrich, J.J. (1970). Convergence lines in southern California. *Weather*, **25**, 140–6.

Andersson, T. & Lindgren, B. (1992). A sea-breeze front seen by radar. *Meteorol. Mag.*, **121**, 230–41.

Angell, J.K. & Pack, D.H. (1965). A study of the sea breeze at Atlantic City, New Jersey using tetroons as lagrangian tracers. *Mon. Weather Rev.*, **93**, 475–93.

Appleton, W.C. (1892). The sea-breeze at Cohasset, Mass. *Am. Meteorol. J.*, **9**, 134–8.

Atkinson, B.W. (1981). *Meso-scale Atmospheric Circulations*. Academic Press. 495 pp.

Atlas, D. (1960). Radar detection of the sea breeze. *J. Meteorol.*, **17**, 244–58.

Atmanathan, S. (1931). The katabatic winds of Poona. *Indian Meteorol. Dept. Sci. Notes*, **4**, 101–11.

Austen, Jane (1814). *Mansfield Park*, J.M. Dent Everyman Edition, 1963.

Banerjee, A.K., Chowdhury, A. & Bhattacharjee, T.H. (1975). On deep inland penetration of sea breeze. *Indian J. Meteorol. Hydrol. Geophys.*, **26**, 501–5.

Barlow, E.W. (1927). The meteorology of solar eclipses. *Q.J.R. Meteorol. Soc.*, **53**, 1–24.

Barry, R.G. (1981). *Mountain Weather and Climate*. Methuen. 311 pp.

Bennett, E.W. & Oded, B.-D. (1967). Application of lidar to air pollution measurements. *J. Appl. Meteorol.*, **6**, 500–15.

Bennett, R.C. & List, R. (1975). Lake breeze detection with an acoustic radar. In *16th Radar Meteorological Conference, Houston*, pp. 260–2. American Meteorological Society.

Berry, R.E. & Taylor, L.R. (1968). High-altitude migration of aphids in maritime and continental climates. *J. Anim. Ecol.*, **37**, 713–22.

Biggs, W.G. & Graves, M.E. (1962). A lake breeze index. *J. Appl. Meteorol.*, **1**, 474–80.

Bilham, E.G. (1934). The sea-breeze as a climatic factor. *J. State Med.*, **42**, 40–50.

Black, J.F. & Tarny, B.L. (1963). The use of asphalt coating to increase rainfall. *J. Appl. Meteorol.*, **2**, 557–64.

Blumen, W. (1984). An observational study of instability and turbulence in nighttime drainage flows. *Boundary Layer Meteorol.*, **28**, 245–69.

Boyd, J.G. (1965). Observation of two intersecting radar fine lines. *Mon. Weather Rev.*, **93**, 188.

Bradbury, T. (1989). *Meteorology in Flight*. London: A. & C. Black. 180pp.

Bradbury, T. (1990). Big cumulus lines. *Sailplane & Gliding*, **41**, 122–4.

Brittain, O.W. (1966). A method of forecasting the sea breeze at Manby. *Met. Office Forecasting Techniques Branch, Memo. No. 12*, pp. 9–16.

Britter, R.E. & Simpson, J.E. (1978). Experiments on the dynamics of a gravity current head. *J. Fluid Mech.*, **88**, 223–40.

Byers, H.R. & Rodebush, H.R. (1948). Causes of thunderstorms of the Florida peninsula. *J. Meteorol.*, **5**, 275–80.

Cameron, D. (1980). *Ballooning Handbook*. London: Pelham Books.

Carpenter, K.M. (1979). An experimental forecast using a non-hydrostatic mesoscale model. *Q.J.R. Meteorol. Soc.*, **105**, 629–55.

Chandler, T.J. (1961). *The Climate of London*. London: Hutchinson. 292 pp.

Clarke, R.H. (1955). Some observations and comments on the sea breeze. *Aust. Meteorol. Mag.*, **11**, 47–52.

Clarke, R.H. (1965). Horizontal mesoscale vortices in the atmosphere. *Aust. Meteorol. Mag.*, **50**, 1–25.

Clarke, R.H. (1984). Colliding sea breezes and the creation of internal atmospheric waves: a numerical model *Aust. Meteorol. Mag.*, **32**, 207–26

Cramer, O.P. & Lynott, R.E. (1961). Cross-section analysis in the study of windflow over mountainous terrain. *Bull. Am. Meteorol. Soc.*, **42**, 693–702.

Dampier, W. (1670). *J. Masefield*. London: J. Grant Richards. (1906 edn.)

Davis, W.M., Schultz, L.G. & Ward, R. de C. (1890). An investigation of the sea-breeze. *Ann. Astron. Observ. Harvard Coll.*, **21**(2), 215–65.

Defant, A. (1949). Zur Theorie der Hangwinde. *Arch. Meteorol. Geophys. Bioklimatol. A*, **1**, 421–50.

Defant, F. (1951). Local winds. In *Compendium of Meteorology*, ed T.F. Malone, pp. 658–672. Boston: American Meteorological Society (reprinted 1960).

Dexter, R.V. (1958). The sea-breeze hodograph at Halifax. *Bull. Am. Meteorol. Soc.*, **39**, 241–7.

Diaz, H.F. & Quayle, R.G. (1980). Historical comparisons of coastal air and sea surface temperature. In *2nd Conference on Coastal Meteorology, Los Angeles 1980*, pp. 216–19. American Meteorological Society.

Dickison, R.B.B. (1990). Detection of mesoscale synoptic features associated with spruce budworm moths in eastern Canada. *Phil. Trans. R. Soc. Lond. Ser. B*, **328**, 606–17.

Diver, A. (1973). Sea breeze over central England. *Sailplane & Gliding*, **24**, 38–40.

Dixit C.M. & Nicholson, J.R. (1964). The sea breeze at and near Bombay. *Indian J. Meteorol. Geophys.*, **15**, 603–8.

Drake, V.A. (1982). Insects in the sea-breeze front at Canberra: a radar study. *Weather*, **37**, 134–143.

Eastwood, E. (1967). *Radar Ornithology*. London: Methuen.

Eastwood, E. & Rider, G.C. (1961). A radar observation of a sea breeze front. *Nature*, **189**, 978–80.

Edinger, J.G. & Helvey, R.A. (1961). The San Fernando Convergence Zone. *Bull. Am. Meteorol. Soc.*, **42**, 626–35.

Eggleton, A.E.J. & Atkins, D.H. (1972). *Results of the Tees-side Investigation*. Harwell: Atomic Energy Research Establishment. 177 pp.

Eliot, J. (1900). Observations recorded during the solar eclipse of January 22, 1898 at 154 meteorological stations in India. *Calcutta, Indian Meteorol. Mem.*, **11**(1).

Eloy, J. & Lhoste, F. (1883) Voyages Aeriens, *La Nature, Paris*, **527**, 83–7.

Elsner, J.B., Mecikalski, J.R. & Tsonis, A.A. (1989). A shore-parallel cloud line over Lake Michigan. *Mon. Weather. Rev.*, **117**, 2822–3.

Estoque, M.A. (1961). A theoretical investigation of the sea breeze. *Q.J.R. Meterol. Soc.*, **87**, 136–46.

Estoque, M.A. (1961). A theoretical investigation of the sea breeze. *Q.J.R. Meteorol. Soc.*, 136–46.

Estoque, M.A. (1962). The sea breeze as a function of the prevailing situation. *J. Atmos. Sci.*, **19**, 244–50.

Fergusson, P. (1971). A sea-breeze at Harrogate. *Weather*, **26**, 125–7.

Findlater, J. (1964). The sea-breeze and inland convection: an example of their interrelation. *Meteorol. Mag.*, **93**, 82–9.

Findlater, J. (1971). The strange winds of Ras Asir (formerly Cape Guardafai). *Meteorol. Mag.*, **100**, 46–54.

Fosberg, M.A. & Schroeder, M.J. (1966). Marine air penetration in central California. *J. Appl. Meteorol.*, **5**, 573–89.

Fulton, R., Zrnic, D.S. & Doviak, R.J. (1990). Initiation of a solitary wave family in the demise of a nocturnal density current. *J. Atmos. Sci.*, **47**, 319–37.

Garratt, J.R. & Physick, W.L. (1986). Numerical study of atmospheric gravity currents. 1: simulations and observations of cold fronts. *Beitr. Phys. Atmos.* **59**, 282–300.

Garratt, J.R. & Physick, W.L. (1987). Numerical study of atmospheric gravity currents. 11: evolution and external influences *Beitr. Phys. Atmos.*, **60**, 88–102.

Geotis, S.G. (1964). On sea breeze 'Angels'. In *World Conference on Radio Meteorology, Boulder, Co*, pp 1–4. American Meteorological Society.

Goff, R.C. (1976). Thunderstorm-outflow kinematics and dynamics. *NOAA Tech. Memo. ERL NSSL-75*, 63 pp.

Gregory, P.H. (1961). *The Microbiology of the Atmosphere*. New York: Interscience. 251 pp.

Grenander, S. (1912). *Der Seebrise an der Schwedischen Ostkuste*. Inaugural Dissertation. Uppsala: University of Uppsala. 96 pp.

Gusten, H. & Heinrich, G. (1989). The effect of sea breeze on the photochemical smog level in Athens, Greece. *Special Environmental Report No. 17*, pp. 61–4. Geneva: WMO.

Hallett, S.I. & Samizay, R. (1980). *Traditional Architecture of Afghanistan*. New York & London: Garland Publishing.

Haurwitz, B. (1947). Comments on the sea-breeze circulation. *J. Meteorol.*, **4**, 1–8.

Haydon, F.S. (1941). *Aeronautics in the Union and Confederance Armies*. Baltimore: Johns Hopkins.

Hürliman, M. (1927). *Indien*. Berlin: Verlag Enst Vasmuth.

Holmes, D.A. (1972). Sea breezes in West Pakistan. *Weather*, **27**, 91–2.

Hsu, S.-A. (1970). Coastal air-circulation system: observations and empirical model. *Mon. Weather Rev.*, **98**, 487–509.

Intieri, J.M., Bedard, A.J. Jr, & Hardesty, R.M. (1990). Details of colliding thunderstorm outflows as observed by doppler lidar. *J. Atmos. Sci.*, **47**, 1081–98.

Intieri, J.M., Little, C.G., Shaw, W.J., Banta, R.M., Dunkee, P.A. & Hardesty, R.M. (1990). The land/sea breeze experiment (LASBEX). *Bull. Am. Meteorol. Soc.*, **71**, 656–64.

Jansa, J.M. & Jaume, E. (1946). The sea-breeze regime in the Majorca Island. (In Spanish.) *Rev. Geofis.*, **V**(19), 304–28.

Jeffreys, H. (1922). On the dynamics of wind. *Q.J.R. Meteorol. Soc.*, **48**, 29–46.

Johnson, R.H. *et al.* (1984). Mesoscale effects of various snow cover over north east Colorado. *Mon. Weather Rev.*, **112**, 1141–52.

Kessler, R.C., Eppel, D., Pielke, R.A. & McQueen, J. (1985). A numerical study of the effects of a large sandbar upon sea breeze development. *Arch. Meteorol. Geophys. Bioclimatol. Ser. A*, **34**, 3–26.

Keulegan, G.H. (1957). An experimental study of the motion of salt water from locks into fresh water channels. *US. Natl. Bur. Stand. Rep.* 5168.

Kimura, R. & Eguchi, T. (1978). On dynamic processes of sea- and land-breeze circulation, *J. Meteorol. Soc. Jap.*, **40**, 67–84.

Kobayasi, T., Sasaki, T, & Osanai, T. (1937). Theoretical and experimental studies of convectional circulation and its relation to land and sea-breezes. *Rep. Aeronaut. Res. Inst., Tokyo*, **12**(1), 67 pp.

Koch, S.E., Dorian, P.B., Melfi, S.H., Skillman, W.C. & Whiteman, D. (1991). Structure of an internal bore and dissipating gravity current as revealed by raman lidar. *Mon. Weather. Rev.*, **119**, 857–87.

Koschmieder, H. (1936). Danziger Seewindstudien, I. *Dan. Meteorol. Forsch.*, **8**, 45 pp.

Kot, S.C. & Simpson, J.E. (1987). Laboratory experiments on two crossing flows. In *Proceedings of the 1st Conference on Fluid Mechanics, Peking, 1987*. pp. 731–6. Beijing University Press.

Kurita, H, Ueda, H & Mitsumoto, S. (1990). Combination of local gradient wind systems under light gradient wind conditions and its contribution to the long-range transport of air pollution. *J. Appl. Meteorol.*, **29**, 131–348.

Kusuda, M. & Alpert, P. (1983). Anti-clockwise rotation of the wind hodograph. Part 1. Theoretical study. *J. Atmos. Sci.*, **40**, 487–99.

Kuttner, J. (1949). Periodische Luftlawinen. *Meteorol. Rundsch.*, **2**, 183–4.

Lack, D. (1956). *Swifts in a Tower*. London: Methuen.

Lamb, H.H. (1955). Malta's sea breezes. *Weather*, **10**, 256–64.

Lambert, S. (1974). High resolution numerical study of the sea-breeze front. *Atmosphere*, **12**, 97–105.

Lambie, J.H. (1963). Southern California shear lines. *Aero Rev. (Swiss), Zurich*, No. 2, 91–4.

Latham, R. (1958). *Marco Polo: The Travels*. (transl.) Penguin Classics.

Lee, D.O. (1992). Urban warming? An analysis of recent trends in London's heat island. *Weather*, **47**, 50–6.

Leighton, P.A. (1961). *Photochemistry of Air Pollution*. Academic Press.

Lied, N.T. (1964). Stationary hydraulic jump in a katabatic flow near Davis, Antarctica, 1961. *Aust. Meteorol. Mag.*, **47**, 40–51.

Linden, P.F. & Simpson, J.E. (1986). Gravity-driven flows in a turbulent fluid. *J. Fluid Mech.*, **172**, 481–97.

Longden, P.C. (1987) Weed beet: past, present and future. In *International Sugar Economic Year Book and Directory*. pp. F5–F15. Ratzeburg, Germany: F.O. Licht.

Longden, P.C. Scott, R.K. & Tydlesley, J.B. (1975). Bolting of sugar beet grown in England. *Outlook on Agriculture*, **8**, 188–93.

Lyons, W.A. & Olsson, L.E. (1972). The climatology and prediction of the Chicago lake breeze. *J. Appl. Meteorol.*, **11**, 1254–72.

Lyons, W.A. & Olsson, L.E. (1972a). Detailed mesoscale air pollution transport in the Chicago lake breeze. *Mon. Weather Rev.*, **101**, 387–403.

Mahrer, Y. & Pielke, R.A. (1977). The effects of topography on sea and land breezes in a two-dimensional numerical model. *Mon. Weather Rev.*, **105**, 1151–62.

Mahrer, Y. & Segal, M. (1985) On the effects of islands' geometry and size on inducing sea-breeze circulation. *J. Atmos. Sci.*, **113**, 170–4.

Mahrt, L. & Larsen, S. (1982). Small scale drainage flow. *Tellus*, **34**, 579–87.

Mak, M.K. & Walsh, J.E. (1976). On the relative strengths of the sea and land breezes. *J. Atmos. Sci.*, **33**, 242–50.

Manins, P.C. & Sawford, B.L. (1979). Katabatic winds: A field case study. *Q.J.R. Meteorol. Soc.*, **105**, 1011–25.

Mansfield, D.A., Milford, J.R. & Purdie, P.G.H. (1974). The use of a powered glider as a sensor of meso-scale vertical air motions. *AIAA paper No. 74–1006*, 1–7.

Marshall, B. & Woodward, F.I. (1985). *Instrumentation for Environmental Physiology*. Cambridge University Press.

McAllister, L.G., Pollard, J.R., Mahoney, A.R. & Shaw, P.J.R. (1969). Acoustic sounding: a new approach to the study of atmospheric structure. *Proc. IEEE*, **57**, 579–85.

McPherson, R.D. (1970). A numerical study of the effect of a coastal irregularity on the sea breeze. *J. Appl. Meteorol.*, **9**, 767–77.

Meadows, R.W. (1967) The Chilbolton 25 m steerable aerial. *IEE Elec. Power*, **13**, 193–5.

Meesters, A.G.C.A., Vugts, H.F., van Delden A.J. & Cannemeijer, F. (1989). Diurnal variation of the surface wind in a tidal area. *Beitr. Phys. Atmos.*, **62**, 258–64.

Met. Office (1943). Synoptic division. *Technical Memo No. 58.*

Meyer, J.H. (1971). Radar observations of land breeze fronts. *J. Appl. Meteorol.*, **10**, 1224–32.

Milford, J.R. & Simpson, J.E. (1972). A shearline investigation with an instrumented glider. *Weather*, **27**, 462–4.

Mitsumoto, S. & Ueda, H. (1983). A laboratory experiment on the dynamics of the land and sea-breeze. *J. Atmos. Sci.*, **40**, 1228–40.

Moroz, W.J. & Hewson, E.W. (1966). The mesoscale interaction of a lake breeze and low level outflow from a thunderstorm. *J. Appl. Meteorol.*, **5**, 148–55.

Nakane, H. & Sasano, Y. (1986). Structure of sea-breeze front revealed by scanning lidar observation. *J. Meteorol. Soc. Jap.*, **64**, 787–92.

Narashimha, R., Pradhu, A., Narahari Rao, K. & Prasad, C.R. (1982). Atmospheric boundary layer experiment. *Proc. Indian Natl. Sci. Acad.*, **48A** (Suppl. 3), 173–86.

Neumann, J. & Mahrer, Y. (1971). A theoretical study of the land and sea breeze circulation. *J. Atmos. Sci.*, **28**, 532–42.

Neumann, J. & Mahrer, Y. (1974). A theoretical study of the land and sea breezes of circular islands. *J. Atmos. Sci.*, **31**, 2027–39.

Niino, H. (1987). On the linear theory of land and sea breeze circulation. *J. Meteorol. Soc. Jap.*, **65**, 901–21.

'Notizen' (1894). 2. Hohe der Seebrise. *Ann. Hydrol. Mar. Meteorol.*, **22**, 313–14

Pack, D.H. & Angell, J.K. (1963). A preliminary study of air trajectories in the Los Angeles basin as derived from tetroon flights. *Mon. Weather Rev.*, **91**, 583–604.

Passarelli, R.E., Jr. & Braham, R.R., Jr. (1981). The role of the winter land breeze in the formation of Great Lake snow storms. *Bull. Am. Meteorol. Soc.*, **62**, 482–91.

Pearce, R.P. (1955). The calculation of a sea breeze circulation in terms of the differential heating across the coastline. *Q.J.R. Meteorol. Soc.*, **81**, 351–81.

Pedgley, D.E. (1990). Concentration of flying insects by the wind. *Phil Trans. R. Soc. Lond. Ser. B*, **328**, 631–53.

Pedgley, D.E., Reynolds, D.R., Riley, J.R. & Tucker, M.R. (1982). Flying insects reveal small-scale wind systems. *Weather*, **37**, 295–306.

Pepperdine, E.C. (1966). Interim report on sea-breeze investigation at Scampton (near Lincoln). *Met. Office Forecasting Techniques Branch Memo.*, **12**, 17–24.

Physick, W.L. (1980). Numerical experiments on the inland penetration of the sea breeze. *Q.J.R. Meteorol. Soc.*, **106**, 735–46.

Physick, W.L. & Byron-Scott R.A.D. (1977). Observations of the sea breeze in the vicinity of a gulf. *Weather*, **32**, 373–81.

Pielke, R.A. (1974). A comparison of three-dimensional and two-dimensional numerical predictions of sea breezes. *J. Atmos. Sci.*, **31**, 1577–85.

Plutach (1892). *1st century A.D. Lives.* Transl. J. & W. Langthorne London: A.J. Valby.

Rainey, R.C. (1969). Effects of atmospheric conditions on insect movement. *Q.J.R. Meteorol. Soc.*, **95**, 424–34.

Rainey, R.C. (1976). Flight behaviour and features of the atmospheric environment. In *Insect Flight, Sympasium of the Royal Entomological Society, London*. Vol. 5, pp. 75–112. London: Blackwell.

Ramball, H.H. & Ferguson, S.P. (1919). Influence of the solar eclipse of June 8 1918, upon radiation and other meteorological elements. *Mon. Weather. Rev.*, **47**, 5–16.

Ramis, C. & Alonso, S. (1988). Sea-breeze convergence line on Majorca: a satellite observation. *Weather*, **43**, 288–93.

Rao, D.V. (1955). The speed and other features of the sea-breeze front at Madras. *Indian J. Meteorol. Geophys.*, **6**, 233–42.

Raynor, G.S., Hayes, J.V. & Ogden, E.C. (1974) Mesoscale transport and dispersion of airborne pollens. *J. Appl. Meteorol.*, **13**, 87–95.

Reible, D.D., Simpson, J.E. & Linden, P.F. (1993). The sea breeze and gravity current frontogenesis. *J.R. Meteorol. Soc.*, **119**, 1–16.

Rider, G.C. & Simpson, J.E. (1968). Two crossing fronts on radar. *Meteorol. Mag.*. **97**, 24–30.

Roaf, S. (1992). *The Wind-catchers of the Middle East*. Wisbech, Cambs: Menas Press.

Rolt, L.T.C. (1966). *The Aeronauts*. London: Longmans.

Rossi, V. (1957). *Land- und Seewind an der finnischen Kusten*. Mitteilungen der Meteorologischen Zentralanstalt, No. 41. Helsinki. (In German.)

Rottman, J.W. & Simpson, J.E. (1989). The formation of internal bores in the atmosphere: a laboratory model. *Q.J.R. Meteorol. Soc.*, **115**, 941–63.

Rotunno, R. (1983). On the linear theory of land and sea breeze. *J. Atmos. Sci.*, **40**, 1999–2009.

Ryznar, E. & Touma, J.S. (1981). Characteristics of true lake breezes along the eastern shore of Lake Michigan. *Atmos. Environ.*, **15**, 1201–5.

Scaetta, H. (1935). Les avalanches d'air dans les Alpes et dans les hautes montagnes de l'Afrique centrale. *Ciel et Terre*, **51**, 79–80.

Schaefer, G.W. (1979). An airborne radar technique for the investigation and control of migrating pest species. *Phil. Trans. R. Soc. Lond. Ser. B*, **287**, 459–65.

Schmidt, F.H. (1947). An elementary theory of the land- and sea-breeze circulation. *J. Meteorol.*, **4**, 9–15.

Schmidt, W. (1911). Zur Mechanik der Boen. *Z. Meteorol.*, **28**, 355–62.

Scorer, R.S. (1990). *Satellite as Microscope*. Chichester: Ellis Horwood.

Segal, M. & Arritt, R.W. (1992). Nonclassical mesoscale circulations caused by surface sensible heat-flux gradients. *Bull. Am. Meteorol. Soc.*, **73**, 1593–604.

Segal, M., Cramer, J.H., Pielke, R.A., Garratt, J.R. & Hildebrand, P. (1991). Observational evaluation of the snow breeze. *Mon. Weather. Rev.*, **119**, 412–24.

Seinfeld, J.H. (1975). *Air pollution: Physical and Chemical Fundamentals*. New York: McGraw Hill.

Shimizu, H, Sasano, Y., Nakane, H., Sugimoto, N., Matsui, I. & Takeuchi, N. (1985). Large scale laser radar for measuring aerosol distribution over a wide area. *Appl. Opt.*, **24**, 617–26.

Simpson, J.E. (1964). Sea-breeze fronts in Hampshire. *Weather*, **19**, 208–19.

Simpson, J.E. (1966). The sea-breeze at Lasham. In *On Sea Breeze Forecasting Techniques*. Memo No. 12, Forecasting Techniques Branch. Met. Office Bracknell.

Simpson, J.E. (1967). Swifts in sea-breeze fronts. *Brit. Birds*, **60**, 225–39.

Simpson, J.E. (1968). A film: 'Sea-breeze fronts'. *Weather*, **23**, 246–7.

Simpson, J.E. (1972). Effects of the lower boundary on the head of a gravity current. *J. Fluid Mech.*, **53**, 759–68.

Simpson, J.E. (1974). A sea-breeze moves seawards. *14th OSTIV Congress, Wailerie, Australia*. Vol 12 (1974), p. 737; vol. 1 (1975) pp. 47–50. Switzerland: Aero Revue.

Simpson, J.E. (1987). *Gravity currents: In the Environment and the Laboratory*. Chichester: Ellis Horwood. 244 pp.

Simpson, J.E. & Britter R.E. (1979). The dynamics of the head of a gravity current advancing over a horizontal surface. *J. Fluid Mech.*, **94**, 477–93.

Simpson, J.E. & Britter R.E. (1980). A laboratory model of an atmospheric mesofront. *Q.J.R. Meteorol. Soc.*, **106**, 485–500.

Simpson, J.E. & Linden, P.F. (1989). Frontogenesis in a fluid with horizontal density gradients. *J. Fluid Mech.*, **202**, 1–16.

Simpson, J.E., Mansfield D.S. & Milford, J.R. (1977). Inland penetration of sea-breeze fronts. *Q.J.R. Meteorol. Soc.*, **103**, 47–76.

Smith, R.A. (1987). Aerial observations of the Yugoslavian Bora. *J. Atmos. Sci.*, **44**, 269–97.

Smith, R.K. & Goodfield, J. (1981). The 1979 Morning Glory Expedition. *Weather*, **36**, 130–6.

Staley, D.O. (1957). The low-level sea breeze of Northwest Washington. *J. Meteorol.*, **14**, 458–70.

Stevens, E.R. (1975). Chemistry and meteorology in an air pollution episode. *J. Air Poll. Contr. Ass.*, **25**, 271–4.

Stevenson, R.E. (1969). The ocean and its atmosphere as seen from Gemini spacecraft. *Ann. Meteorol., Offenbach*, **4**, 195–9.

Streten, N.A. (1963). Some observations of antarctic katabatic winds. *Aust. Meteorol. Mag.*, **42**, 1–23.

Thompson, B.W. (1984). Small-scale katabatics and cold hollows. *Weather*, **41**, 146–53.

Tissandier, G. (1871). Travels of MM Fonvielle and Tissandier. In *Travels in the Air*, J. Glaisher, pp. 233–395 London: Richard Bentley.

Tsuboki, K., Fujiyoshi, Y. & Wakakama, G. (1989). Structure of a land breeze and snowfall enhancement at the leading edge. *J. Meteorol. Soc. Jap.*, **67**, 757–70.

Tyson, P.D. (1968). Velocity fluctuations in the mountain wind. *J. Atmos. Phys.*, **23**, 381–4.

Ueda, H., Mitsumoto, S. & Kurita, H. (1988). Flow mechanism for the long-range transport of air pollutants by the sea breeze causing inland nighttime high oxidants. *J. Appl. Meteorol.*, **27**, 182–7.

Wallington, C.E. (1961). *Meteorology for Glider Pilots*. London: John Murray, 279 pp.

Walsh, J.E. (1974). Sea breeze theory and applications. *J. Atmos. Sci.*, **31**, 2012–26.

Waters, A.J. (1990). Radar image: 19 October 1990 at 0600 UTC. *Meteorol. Mag. London*, **119**, 272.

Watts, A. (1955). Sea breeze at Thorney Island. *Meteorol. Mag.*, **84**, 42–8.

Watts, A. (1965). *Wind and Sailing Boats*. London: Adlard Coles.

Wirth, D. & Young, J. (1980). *Ballooning*. London: Orbis Publishing.

Wood, I.R. & Simpson, J.E. (1984). Jumps in layered miscible fluids. *J. Fluid Mech.*, **140**, 329–42.

Xian, Z. & Pielke, R.A. (1991). The effects of width of land masses on the development of sea breezes. *J. Appl. Meteorol.*, **30**, 1280–1304.

Yan, B. & Huang, R. (1988). A numerical study of the land and river breezes over Chongqing on the Yangtse River. *Chin. J. Atmos. Sci.*, **13**, 159–70

Yoshino, M.M. (ed.) (1976). *Local Wind Bora*. Tokyo: University of Tokyo Press.

Index